# Solar Photovoltaics: Business Briefing

David Thorpe

News editor, *Energy and Environmental Management* and author of *Solar Technology* (Earthscan, 2011)

First published in 2012 by Dō Sustainability
87 Lonsdale Road, Oxford OX2 7ET, UK



ISBN 978-1-909293-04-5 (eBook-ePub)
ISBN 978-1-909293-05-2 (eBook-PDF)
ISBN 978-1-909293-03-8 (Paperback)

A catalogue record for this title is available from the British Library.

At Dō Sustainability we strive to minimize our environmental impacts and carbon footprint through reducing waste, recycling and offsetting our $CO_2$ emissions, including those created through publication of this book. For more information on our environmental policy see **www.dosustainability.com.**

Page design and typesetting by Alison Rayner
Cover by Becky Chilcott

For further information on Dō Sustainability, visit our website:
**www.dosustainability.com**

# DōShorts

**Dō Sustainability** is the publisher of **DōShorts**: short, high-value ebooks that distil sustainability best practice and business insights for busy, results-driven professionals. Each DōShort can be read in 90 minutes.

## New and forthcoming DōShorts -- stay up to date

We publish 3 to 5 new DōShorts each month. The best way to keep up to date? Sign up to our short, monthly newsletter. Go to **www.dosustainability.com/newsletter** to sign up to the Dō Newsletter. Some of our latest and forthcoming titles include:

- ***Green Jujitsu: Embed Sustainability into Your Organisation*** Gareth Kane
- ***How to Make your Company a Recognised Sustainability Champion*** Brendan May
- ***Making the Most of Standards*** Adrian Henriques
- ***Promoting Sustainable Behaviour: A Practical Guide to What Works*** Adam Corner
- ***Solar Photovoltaics Business Briefing*** David Thorpe
- ***Sustainability in the Public Sector*** Sonja Powell
- ***Sustainability Reporting for SMEs*** Elaine Cohen
- ***Sustainable Transport Fuels Business Briefing*** David Thorpe
- ***The Changing Profile of Corporate Climate Change Risk*** Mark Trexler & Laura Kosloff
- ***The First 100 Days: Plan, Prioritise & Build a Sustainable Organisation*** Anne Augustine
- ***The Short Guide to SRI*** Cary Krosinsky

## Subscriptions

In additional to individual sales and rentals, we offer organisational subscriptions to our full collection of published and forthcoming books. To discuss a subscription for your organisation, email **veruschka@dosustainability.com**

## Write for us, or suggest a DōShort

Please visit **www.dosustainability.com** for our full publishing programme. If you don't find what you need, write for us! Or Suggest a DōShort on our website. We look forward to hearing from you.

# Contents

1 About This Book ..... 11
1.1 **Who it is for** ..... 11
1.2 **Introduction** ..... 11
1.3 **Energy basics** ..... 12
2 What is Solar Power? ..... 15
2.1 **How much energy is there?** ..... 15
2.2 **Solar electricity** ..... 19
***2.2.1 Photovoltaics*** ..... 22
2.2.1.1 Advantages ..... 22
2.2.1.2 Disadvantages ..... 23
***2.2.2 Where best to deploy PV*** ..... 24
2.3 **The growth in PV** ..... 27
***2.3.1 Reality check*** ..... 29
2.4 **Costs are falling** ..... 30
2.5 **Energy efficiency** ..... 32
3 The Photovoltaic Effect ..... 33
3.1 **Types of cell** ..... 33
***3.1.1 Silicon*** ..... 33
3.1.1.1 Concentrated solar power ..... 36
***3.1.2 Thin film modules*** ..... 37

3.2 **Power output** 40
***3.2.1 Lifetime*** 40
3.3 **Environmental impact** 41
***3.3.1 Recycling*** 41
***3.3.2 Energy payback*** 42
4 Applications 45
4.1 **Grid-connected - small-scale** 45
4.2 **Grid-connected - larger or utility scale** 46
***4.2.1 The advantage of speed*** 48
4.3 **Building Integrated PV (BIPV)** 48
4.4 **Stand-alone systems** 49
***4.4.1 Telecoms and weather stations*** 49
***4.4.2 Road furniture*** 50
***4.4.3 Gadgets*** 50
4.5 **Transport** 51
4.6 **PVs in space** 52
5 The Cost of PV Systems 53
***5.0.1 The 'levelised cost of energy' (LCOE)*** 53
***5.0.2 Cost uncertainties*** 54
***5.0.3 Grid parity*** 56
5.1 **Capital costs** 58
***5.1.1 Operating costs*** 61
***5.1.2 Cost comparisons*** 61
5.1.2.2 Location drivers 65

*5.1.2.3 Policy drivers* ... 66
5.2 **Feed-in Tariffs** ... 66
*5.2.1 Capacity trigger* ... 72
5.3 **State of the industry** ... 73
*5.3.1 Long-term outlook* ... 73
*5.3.2 Desertec* ... 74
*5.3.3 Short-term outlook* ... 75
*5.3.4 Outlook for solar farms* ... 76
5.4 **Advice to installers** ... 77
6 Planning a Solar Installation ... 79
6.1 **System components** ... 79
6.2 **Calculating output** ... 79
*6.2.1 UK SAP guidelines* ... 81
6.3 **General design advice** ... 83
6.4 **Where should it go?** ... 84
6.5 **Costs** ... 84
*6.5.1 Estimating cost saving* ... 85
*6.5.2 Estimating $CO_2$ emissions saved* ... 86
6.6 **Inverters** ... 86
6.7 **Maintenance** ... 87
6.8 **Sourcing and talking to suppliers** ... 87
6.9 **Insurance** ... 88
6.10 **Warranties** ... 88
6.11 **Generation data** ... 88

7 Investing in PV ... 91
7.1 **Drivers for PV** ... 92
7.2 **UK Standards** ... 95
7.3 **Planning** ... 96
7.4 **Business plans** ... 97
***7.4.1 Implementation timescale*** ... 98
7.5 **Financial support** ... 98
***7.5.1 Grants and subsidies*** ... 99
***7.5.2 Community projects*** ... 99
***7.5.3 Tax rebates*** ... 99
7.6 **Investment opportunities** ... 100
***7.6.1 The Clean Development Mechanism (CDM)*** ... 100
***7.6.2 Joint Implementation (JI)*** ... 101
***7.6.3 Voluntary Emissions Reductions or Verified Emissions Reductions (VERs)*** ... 102
***7.6.4 Loans to installers*** ... 102
7.7 **Utility level developers** ... 103
***7.7.1 Energy Service Companies (ESCOs)*** ... 103

8 Sources of Information ... 105
8.1 **On official UK policy** ... 105
8.2 **On solar irradiance** ... 107
***8.2.1 MIDAS*** ... 107
***8.2.2 Juice-o-meter*** ... 108
***8.2.3 Photovoltaic Geographical Information System (PVGIS)*** ... 109

8.3 **Advice** ........ 110

Decision Tree ........ 111

About the Author ........ 113

CHAPTER 1

# About This Book

**THIS PUBLICATION OUTLINES**, for a UK business audience: the technical and scientific basis of PV technologies; their applications and how to assess them; the prospects and drivers for cost reductions and implementation; and the business case for and against investment.

## 1.1 Who it is for

This report should be useful to anyone considering a business use of solar PV. This might be an investor; someone considering installing a system; someone considering PV for carbon offset purposes; or someone considering setting up a solar PV business. It should enable the reader to have sufficient knowledge to, for example, either talk to contractors, or to set about making a business case for investment to senior management.

## 1.2 Introduction

Solar photovoltaics (PV) is the sunrise sector for electricity generation – the renewable technology whose time has come. Clean and with no moving parts to wear out, they interface neatly with other technologies in our digital world. Cost curves are decreasing and installation curves exponentially rising over a three-decade timescale.

Although silicon-based cells are already well-known, due to the Feed-in Tariffs support they receive, within the next five to eight years lowering production costs and technological innovations will mean that solar electricity will be poised to find even more widespread applications. Some of these will be distributed grid-feeding installations on many new and retrofit buildings in the business sector in the UK, others will be niche applications; at a utility scale, most will be in the sunniest parts of the world.

## 1.3 **Energy basics**

It's helpful to have a basic understanding on energy and power.

- The amount of electricity in a circuit is measured in volts.
- The force of the electrical flow, or current, is measured in amps.
- The power in a circuit is defined by volts times amps, and called a watt. W = V x A.
- A watt (W) is therefore the unit of power or the rate at which work is done when one ampere (A) of current flows through an electrical potential difference of one volt (V).
- 1000 watts is a kilowatt (kW).
- Energy is the amount of power produced by a generator or consumed by an appliance or over a period of time. Its unit is the watt-hour (Wh).

- 1000 watt-hours is a kilowatt-hour (kWh), commonly a unit of electricity on an electricity bill. It is the rate of generation or consumption of power.
- An alternative unit for this is the joule (J). 3600 joules = 1 Wh. A joule is also one watt per second, since there are 3600 seconds in an hour; or 3.6 megajoules (MJ) = 1kWh.

For example:

- One PV solar panel producing 80W for two hours, or two panels producing 80 W for one hour would produce 2 x 80 = 160 Wh.
- Three panels producing 90 W for five hours will produce 3 x 90 x 5 = 1350 Wh or 1.35 kWh.

.....................................................................................................

CHAPTER 2

# What is Solar Power?

**ALL LIFE ON EARTH** owes its existence to the energy coming from the sun. It powers the weather and hence all forms of renewable energy. Plants convert its energy using photosynthesis into biomass, and all fossil fuels are derived from prehistoric biomass.

## 2.1 How much energy is there?

Books on solar energy often say something like: 'In less than two hours, enough energy reaches the Earth's surface from the sun to satisfy humanity's current energy needs for a whole year.' But of course we can't use all that solar energy. The real question is: how much can we use, and where?

The answer is highly site-specific, to do with the latitude, climate and much more. The edge of the Earth's atmosphere receives 174 petawatts (PW) of solar radiation (or insolation), otherwise written as $1.74 \times 10^{17}$ W. This figure is called the 'solar constant'. About half of this reaches the Earth's surface; 5800 times humanity's average energy consumption rate of 15 terawatts ($1.5 \times 10^{13}$ W).

**FIGURE 2.1a. Solar irradiation and where it goes within the Earth's ecosystem.**

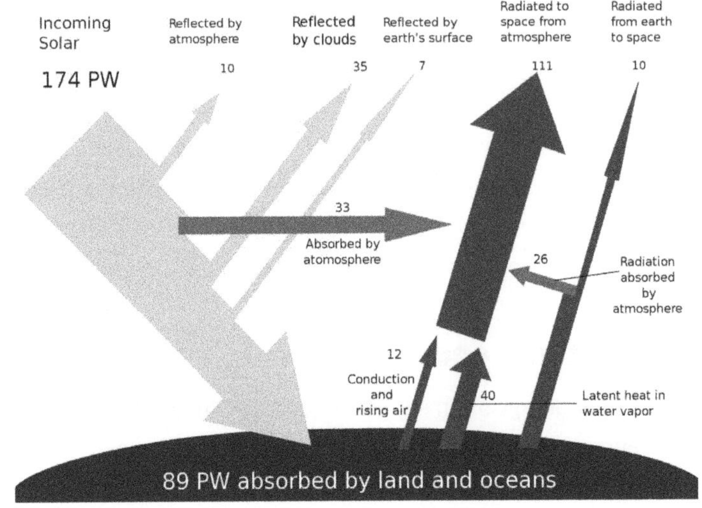


How much each location receives varies over the year because of the tilt of the Earth's axis. This gives rise to the seasons, as first one hemisphere and then the other receives more solar energy.

**FIGURE 2.1b. The effect of the Earth's tilt on surface insolation.**

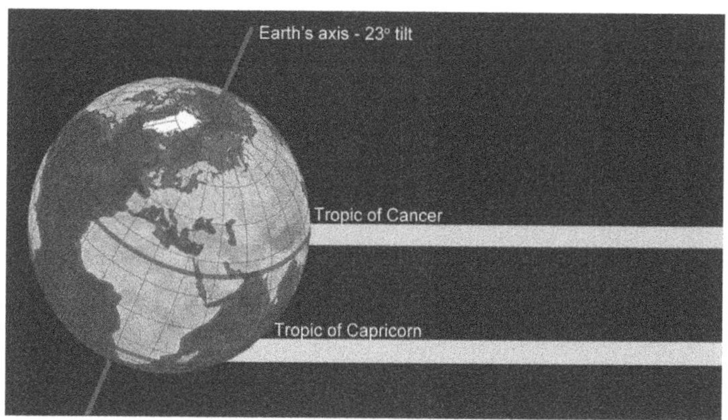


Arid and hot areas, known as the 'sun belt', are best suited to many forms of solar power, because of the direct, year-long sunlight and little cloud cover.

**FIGURE 2.1C. Global insolation averages.**

Figure 10-2 Average yearly solar radiation, mean values 1981-2000
Source: Energie-Atlas GmbH

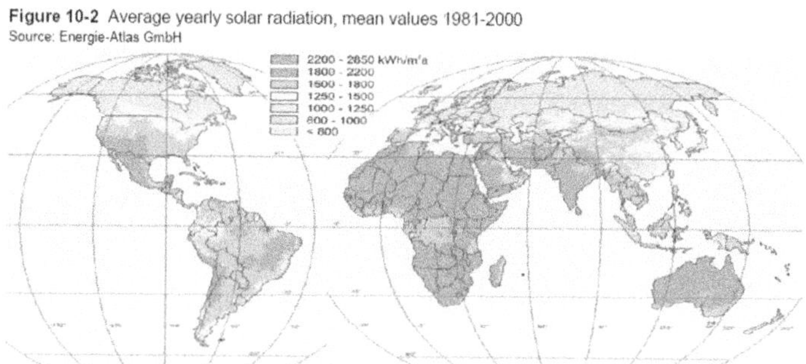


**SOURCE:** Energie-Atlas GmbH.

**FIGURE 2.1d: Andasol solar thermal power station, Spain.**

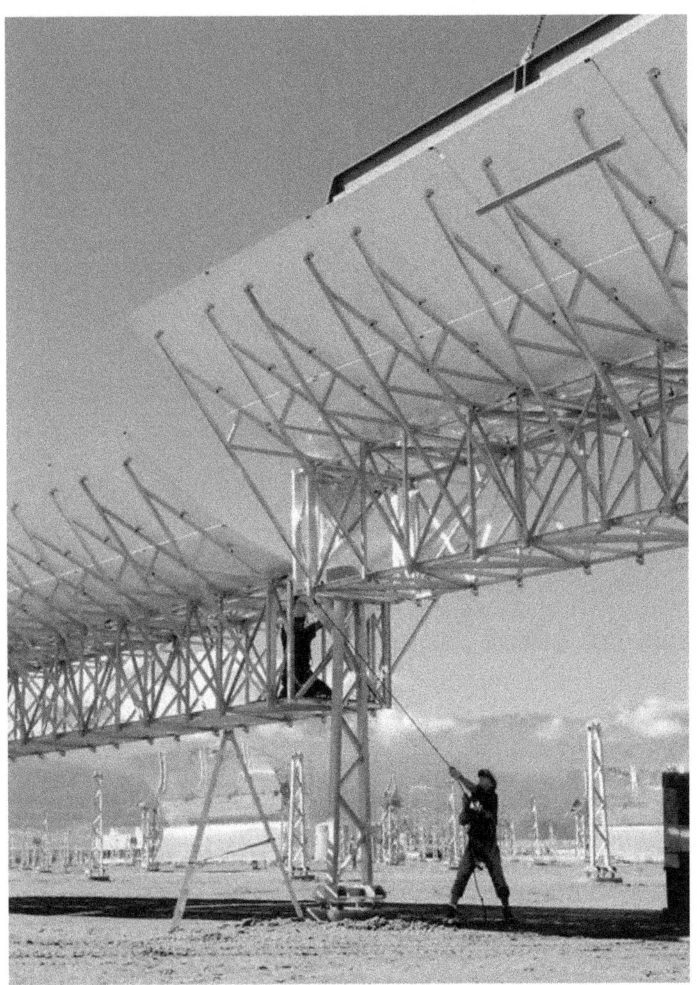

Higher altitude also has advantages because less energy has been lost in the atmosphere. For example, the new €350 million Andasol solar thermal farm in Spain, with 600,000 parabolic sun-tracking mirrors, is located on the Guadix plateau 1100 meters above sea level because it receives 2000 hours of sunlight per year.

Electricity is not the only way to use the energy of the sunshine. It can also be used:

- to heat or cool water (solar thermal)
- to heat or cool buildings (solar architecture, sometimes known as passive solar design)
- or indirectly by exploiting wind power, hydroelectric power, marine energy such as wave and tidal power, burning biomass or using heat pumps. All of these are indirectly driven by the sun's heat either causing thermal currents or being stored in plants or the Earth's mass.

## 2.2 **Solar electricity**

Electricity is the most versatile form of power that we know of. There are several ways of turning sunshine into electricity as well as PV, some of which use the sun's light wavelengths and some of which use its heat (thermal wavelengths). The principal technologies are:

- photovoltaic (PV) (using its light);
- concentrating solar thermal power (CSP) (using its heat);
- solar towers (using its heat);
- thermoelectrics (using its heat).

**FIGURE 2.2a. The wavelengths of light and their applications.**

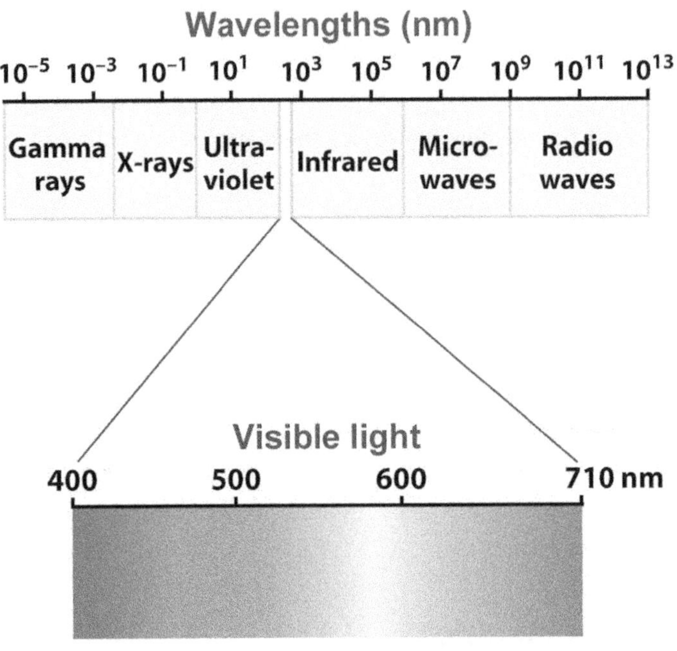


Concentrating solar power uses parabolic dishes or troughs to focus the sun's heat onto a liquid. The liquid boils and the steam drives a conventional generator. This technique has been in existence since the 1880s.

Solar towers are surrounded by an array of mirrors which track the sun and focus its heat onto a point at the top of the tower, where steam is produced which also drives a generator.

**FIGURE 2.2b. Augustin Mouchot's concentrating parabolic dish, used to drive a steam engine and print a newspaper in the 1880s.**

Thermoelectrics use heat, whether from the sun or any other source such as process heat in a factory, to produce an electric current from a thermocouple.

This book is not concerned with the last three, only with photovoltaics.

## 2.2.1 Photovoltaics

The word 'photovoltaic' comes from a Greek word *photos* meaning 'light' and 'volt' – the SI unit of electromotive force, named after its discoverer, Alessandro Volta. PV solar modules generate electricity from the sun's light.

### 2.2.1.1 Advantages

Solar electricity has many advantages:

- low-maintenance, due to no moving parts
- electricity is from a free source
- no pollution or contribution to global warming, so credits may be obtainable to help finance the upfront capital cost
- long-lasting: modules may continue to provide free energy for up to 30 years, yielding a feed-in-tariff income-stream for eligible grid-fed electricity, although electrical output can tail off slightly with age
- scalable: by adding more modules together, the amount of power that can be generated is only limited by the available space and budget.

### 2.2.1.2 Disadvantages

But there are downsides:

- intermittency; it only works when the sun is shining. Therefore, if we want electricity during the night or cloudy periods, we must either store the energy in some form that can be used later, such as in a battery or as hydrogen, or supplement our supply with electricity from the grid.
- seasonality; as latitudes approach the poles solar energy becomes increasingly seasonal, so when we need electricity most, in winter during longer nights, there is less of it available. Therefore at these latitudes, PV systems must be appropriately sized, or supplemented with other sources of electricity.
- deterioration; modules may produce 95% of rated output for the first 10 years, 90% for the next 10 years, then continue to deteriorate. The rate of deterioration depends on the type of cell used. Latest results from modules operating since the 1980s show a degradation rate of 0.2 to 0.5% per year.

## 2.2.2 Where best to deploy PV

Most common types of silicon (Si) PV cells are best suited to direct sunlight, but some (the new dye-sensitised solar cells) work well in diffuse (indirect) light as well as direct light. All cells will generate some electricity at any light level. In Northern Europe, on average over a year, half the total solar radiation is direct and half diffuse.

**FIGURE 2.2.2a. Insolation map for the UK**

The amount of energy varies from 1100 kW/m$^2$ in Cornwall to 750 kW/m$^2$ in the Orkneys, making the South-west the area which would recoup investment in PV in the quickest time.

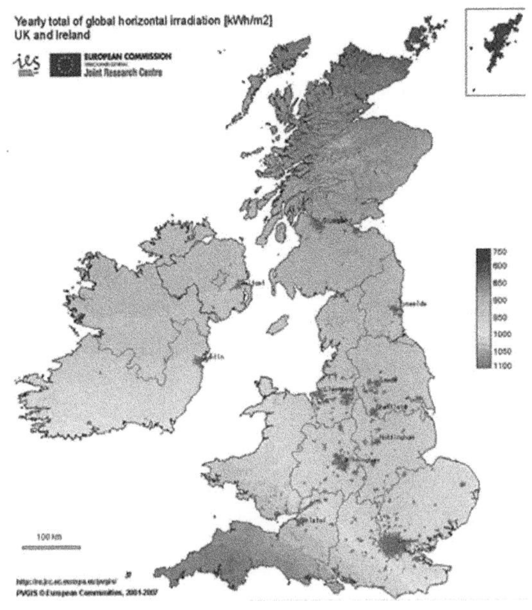

**FIGURE 2.2.2b. The effect of the angle of incidence on the amount of solar energy received: an angle of 30° halves the energy content.**

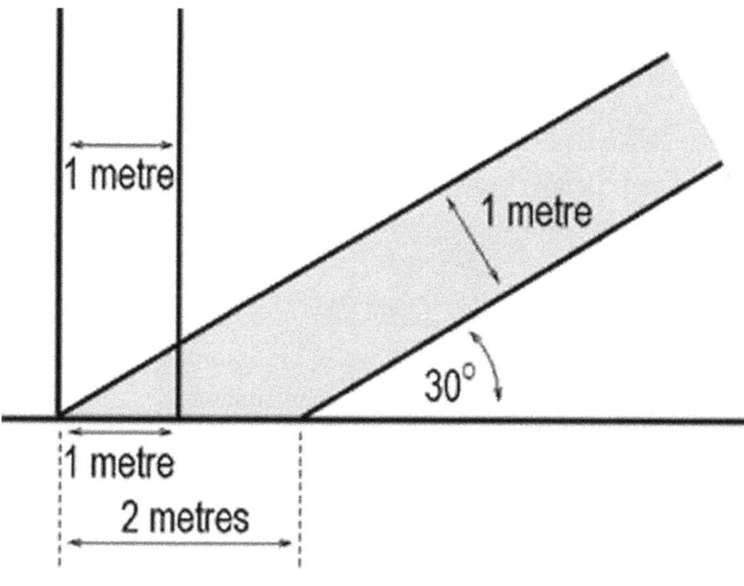


The angle at which the light hits the module is important. Ideally, it needs to hit the module straight on. At 30 degrees from the horizontal, for example, the same amount of light will be spread over double the area of the module, halving its effectiveness.

Utility scale silicon module installations often track the sun across the sky to maintain the optimum angle. If the modules are to be fixed, the optimum angle will be chosen by the contractor, but will always face the equator and usually point towards the sun's average position in the sky at midday. This is called the azimuth angle.

**FIGURE 2.2.2c. The azimuth angle and its variation through the day and seasons.**

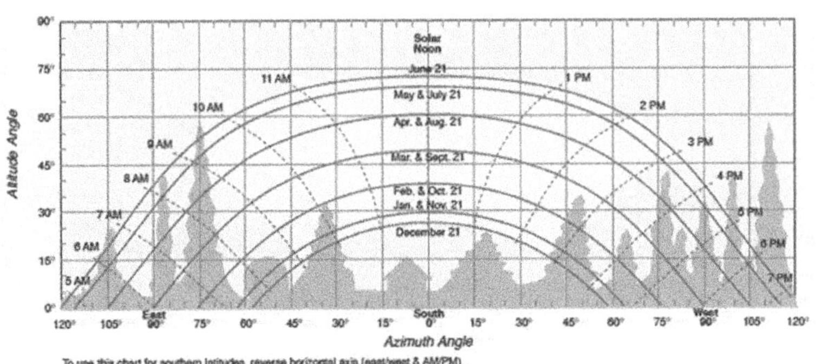


**TABLE 2.2.2. The effect of tilt angle and orientation.**

The table shows that the efficiency of a PV array is compromised by 17% when placed vertically on a building façade, and that east- or west-facing arrays suffer a 50% loss if placed vertically.

**'Retscreen' simulation for high rise in Manchester with no shading. Average Energy Generation**

| Orientation | Tilt | Generation kWh/m²/day | Reduction from optimum |
|---|---|---|---|
| South | 35° | 3.00 | Optimum |
| South | Vertical | 2.18 | 17% |
| South | Horizontal | 2.63 | 12% |
| East/West | 35° | 2.46 | 18% |
| East/West | Vertical | 1.63 | 47% |
| East/West | Horizontal | 2.63 | 12% |

They must also have no shadows falling on them during the main part of the day, for this will also reduce the amount of electricity generated. It follows that the more sunlight there is falling on the module, then the faster the economic payback will be on the cost of the system. But economic factors may not always be the reasons for choosing solar power.

Developers or contractors conduct site surveys using instruments which measure solar irradiation, broken down into several components such as direct, diffuse and reflected light, and feed them into functions that take into account the above variables plus the azimuth. They compare these to both freely and commercially available figures, sources of which can be found in section 8.

## 2.3 **The growth in PV**

Globally, installations of solar PV are happening at a faster rate than any other energy technology. The curve in the graph of Figure 2.3a is almost exponential. Solar power capacity needs to increase, since global energy demand is rising just as quickly as current solar power deployment. Partly depending on political support, PriceWaterhouseCoopers and others estimate that it is possible that over the next 50 years solar PV can meet a much higher percentage of demand and, together with energy efficiency measures and the other renewable energy sources, gradually replace fossil fuels. (**SOURCE:** *100% renewable electricity: A roadmap to 2050 for Europe and North Africa*, PwC, 2010, at **www.pwc.com/sustainability**.)

Figures for 2010, the latest available, show that throughout the world the total installed PV capacity was 40 gigawatts (GW), producing 50 terawatt-hours per year (TWh/yr). (**SOURCE:** EPIA, 2011.) Europe is leading the way in installations, with over 13 GW of PV capacity constructed in 2010;

**FIGURE 2.3a. The growth of global installed capacity of PV 1995–2010.**

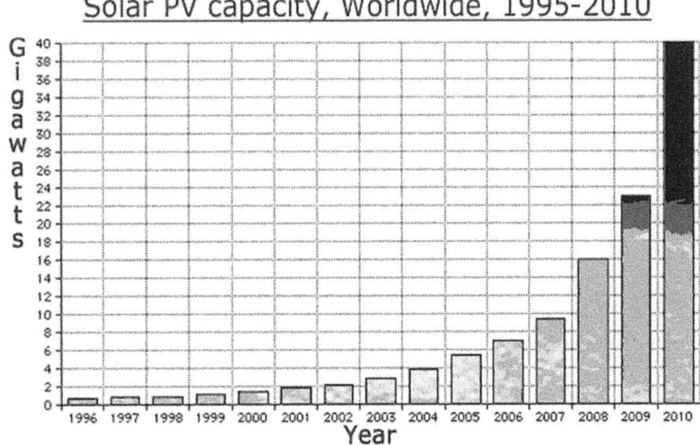


**SOURCE:** Worldwatch Institute/REN21.

the rest of the world installed over 17 GW. Within Europe, Germany, Italy, Greece and Spain lead the way, with Germany having 44% of total world capacity. 0.02GW of Concentrated PV was connected to the grid worldwide during 2010 and early 2011 (see section 3.1.1).

For the first year ever, Europe added more PV than wind capacity during 2010, and this trend continued through 2011. However, the average cost of PV in Europe at a utility scale is still three to four times that of onshore wind and biomass.

Looking at the wider context, renewable energy sources in general see growth rates of up to 70% every year, while fossil fuels' annual growth is in the low digits, and nuclear energy's share is further shrinking. Renewables

**FIGURE 2.3b shows which countries had the most installed capacity of PV in 2010, with Germany the clear leader due to its early adoption of feed-in-tariffs.**

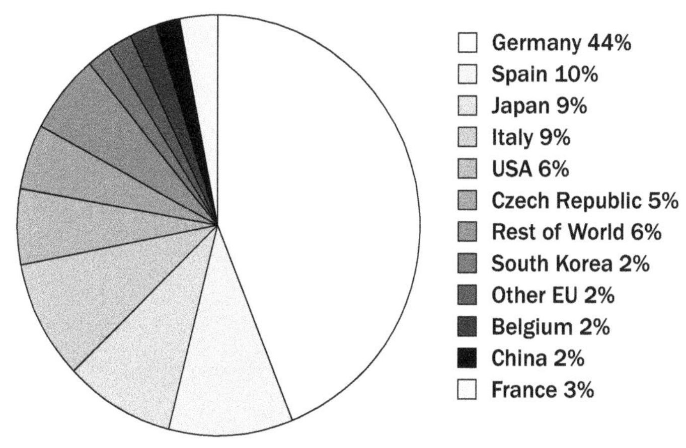


supplied 16% of global primary energy consumption in 2010, and 20% of worldwide electricity. The sector's growth has been a positive constant, despite the economic recession and public finance crises in many parts of the world. Among the most rapidly growing national renewable energy markets in 2010 were world-leader China (+26%), Germany (+10.4%), and the US (+5%). (**SOURCE:** Worldwatch Institute/REN21.)

## 2.3.1 Reality check

And yet, in 2009, PV still generated under 0.2% of total global electricity demand (17.3 trillion kWh in 2008). The International Energy Agency

(IEA) projects that this total demand will rise by 78% to 4.7 TWh by 2030. Therefore, even if PV keeps up its current expansion rate, it will barely keep pace with the increase in overall demand.

To make a real difference, deployment must increase at an even faster rate, which requires strong, international political will, and robust, stable policies and regulation, with a steady supply of capital.

## 2.4 **Costs are falling**

The prospect of this happening is helped by a number of factors, but hindered by the global economic recession and low price of carbon. On the plus side, module prices are falling rapidly. They almost halved in 2010–11, as Table 2.4 and Figure 2.4 show.

**TABLE 2.4. The price of PV in June 2010 and November 2011**

| Module pricing per peak watt | Unit | June 2010 | Nov 2011 | % reduction in 17 months |
|---|---|---|---|---|
| Europe | €/watt | 4.13 | 2.33 | 0.56% |
| US | $/watt | 4.23 | 2.49 | 0.59% |
| Lowest mono-crystalline module price | $/Wp | 2.23 | 1.28 | 0.57% |
| | €/Wp | 1.65 | 0.91 | 0.55% |
| Lowest multi-crystalline module price | $/Wp | 1.74 | 1.31 | 0.75% |
| | €/Wp | 1.29 | 0.93 | 0.72% |
| Lowest thin-film module price | $/Wp | 1.76 | 1.25 | 0.71% |
| | €/Wp | 1.3 | 0.89 | 0.68% |

**FIGURE 2.4. The fall of module prices 2010–2011 compared to PV capacity.**

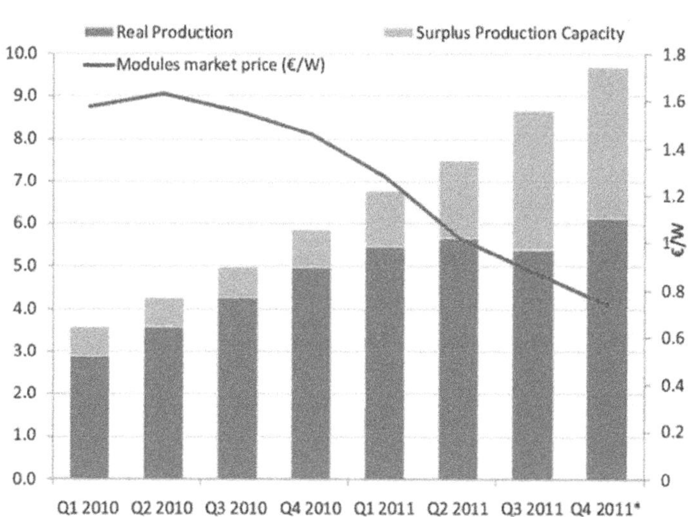


According to the European Photovoltaic Industry Association, 'the price of PV modules decreases by over 20% every time the cumulative sold volume of PV modules has doubled' (EPIA, 2011).

The 50% fall in price of modules in 2011 was due not to technological advances but to a tough competitive environment, due to increased persistent output from well-capitalised Asian manufacturers seeking to gain market share despite their shrinking margins.

As a result, over six months in Germany alone three out of its 12 solar manufacturing companies became insolvent, two were bought up, four terminated their wafer production and only Q-Cells and Solarworld continued their activities but had to reduce their costs and shift production sites.

The cost of modules is only part of the overall system costs. But these too are falling as demand increases. Section 7 looks more at the business implications and drivers for getting involved in PV, and the prospect for subsidies.

## 2.5 **Energy efficiency**

It is always cheaper to save energy and reduce the need for it than it is to generate energy. So every attempt must be made to reduce and manage the demand for electricity before considering installing solar power.

CHAPTER 3

# The Photovoltaic Effect

**THE PHOTOELECTRIC OR PHOTOVOLTAIC EFFECT** was discovered in 1839 by Edmund Becquerel. But it wasn't until 1883 that the first solar cell was built by Charles Fritts. He shone light upon a layer of selenium covered with a layer of gold and generated electricity. The effect was explained by Albert Einstein in 1905 as being due to the quantum nature of light, which can behave like a wave as well as a stream of photons.

The first use of a silicon semiconductor solar cell, made by Bell Laboratories in New Jersey, USA, was to power the second American spacecraft Vanguard 1 in 1958. The most advanced photovoltaic cells are still used to power spacecraft. It is really space flight which has propelled the development of the technology.

## 3.1 **Types of cell**

### 3.1.1 Silicon

Standard crystalline silicon cells (c-Si) are made of two layers of silicon, one of which is 'dosed' with atoms of phosphorous, and the other with boron, to give them opposite electrical charges. Light hitting the upper layer dislodges an electron into the lower layer. If the two are connected in a circuit, a current results.

**FIGURE 3.1.1a. Cross-section through a silicon PV cell.**

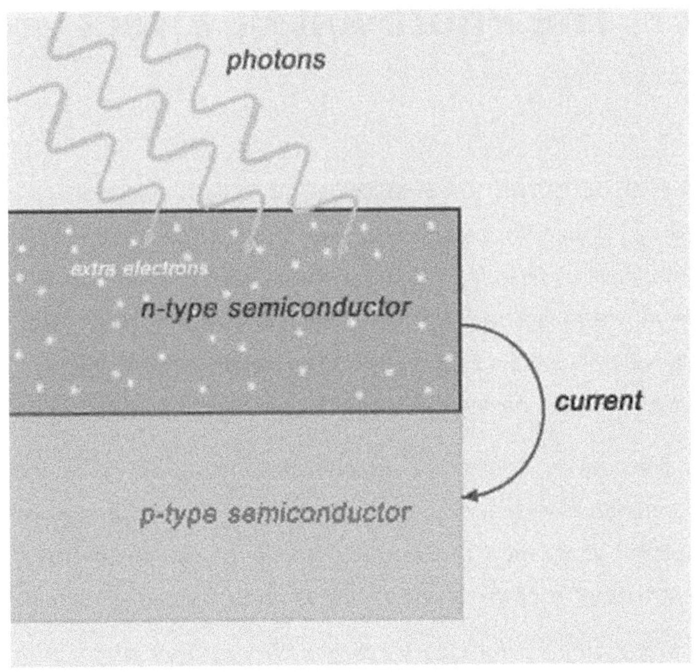


The amount of light energy, and which frequencies of light are able to dislodge the electron, is determined by the 'band gap' of the material. The efficiency of a cell is partly determined by the range of frequencies that can produce this effect. 'Single junction' cells just respond to a narrow frequency range. In more expensive 'multi-junction' cells, different layers have different band gaps, each responding to different frequencies, making the overall cell able to capture more energy from a wider range of frequencies, and therefore produce more current. Higher frequencies possess more energy.

**FIGURE 3.1.1b. Cells are connected together to make modules. Modules may be connected together to make arrays. The amount of power they may generate is the sum of the power output of each cell.**

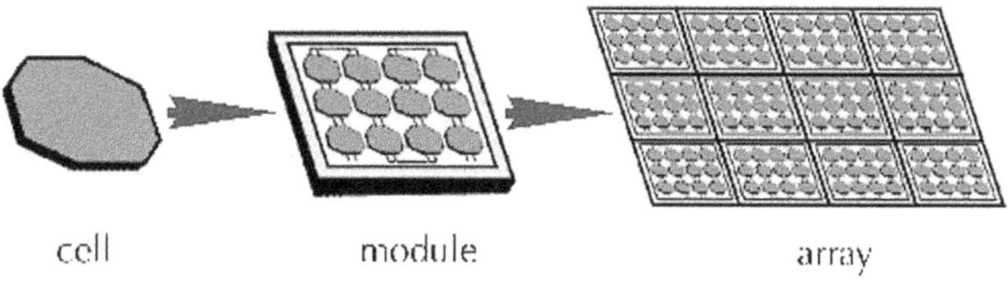


**FIGURE 3.1.1c. From sand to silicon, to a pure block, to a solar panel.**

Silicon cells can be monocrystalline, polycrystalline or amorphous.

- Monocrystalline cells are made from high-quality silicon and can be up to 24% efficient but more expensive.
- Polycrystalline cells are made from silicon that is melted and cast. This is the most common type of cell, representing about 85% of the market.
- Amorphous cells have no crystals and are the cheapest and least efficient, being used in products like garden lights.

Other metals and compounds beside silicon have the same property, and some are cheaper to produce.

### 3.1.1.1 Concentrated solar power

Concentrated photovoltaics (CPV) are a new development. They are used at the moment principally in solar power stations in the sunbelt areas of the world where normal direct irradiation is greater than 1800 kWh/m$^2$/year. They use lenses like magnifying glasses to concentrate sunlight from a wider area onto a smaller area where the PV cell is located. This

**FIGURE 3.1.1.1. Concentrating solar cells are far more efficient because they use lenses to focus light onto the cells.**

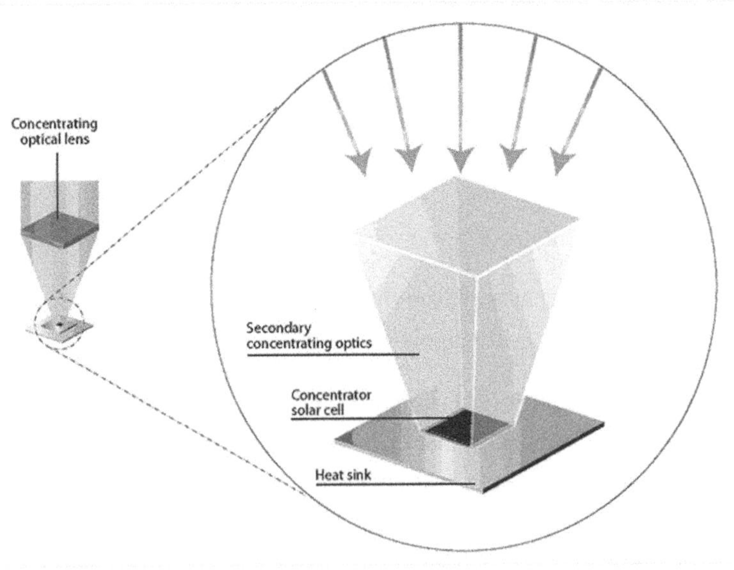

reduces the number of cells required, making an array cheaper, or reduces the amount of land required, for the same amount of power output. They can range from 4x up to 1500x.

Medium-scale concentrators using prismatic lenses, equivalent to around 120 suns, are the most economical. In general, however, concentrating solar thermal electricity-generating power stations are more economical in sunbelt areas.

### 3.1.2 Thin film modules

Thin film modules are a more recent development and have the potential to revolutionise access to electricity because they are so cheap to produce. They consist of extremely thin layers of photosensitive materials printed onto a low-cost backing such as plastic, steel or glass. Whereas typical Si devices have layers 100 mm thick (about the same as a human hair), thin film layers are 1–10 mm thick.

Among the commercialised PV technologies, due to the fact that it is printed, thin film has reached the lowest production costs, even though crystalline silicon (c-Si) module prices are also decreasing. Although cheaper to make than silicon cells, they are less efficient: between 6% and 13%. Therefore, to obtain the same amount of power output, about twice the amount of surface area is required than for polycrystalline modules.

There are three types of thin film modules:

- amorphous silicon (a-Si), as used in calculators, around 6–8% efficient
- cadmium telluride (CdTe), around 10% efficient
- copper indium selenide (CIS)/gallium diselenide/disulphide (CIGS), around 10% efficient.

CdTe is currently the technology with the lowest cost, but issues of scarcity and materials toxicity limit its future growth potential. Tellurium (used in CdTe), indium and gallium do not occur naturally in sufficient quantities to support a mass roll-out of cells utilising them. As a rough guide, 50 tonnes of raw materials are required to manufacture 1 GWp (Gigawatts-peak) worth of cells.

Amorphous silicon (a-Si) thin-film cells are being developed using a variety of different techniques to enable them to trap greater ranges of light frequencies, and hence improve efficiency. R&D is focused on adding more materials in layers to capture more frequencies, and on reducing production costs. If successful at doing this at a mass producible scale, by using around 100 times less material to produce the same amount of

**FIGURE 3.1.2a. A prototype dye-sensitised PV module from Tata Steel and Dyesol.**

electrical output, the cost of modules will be dramatically reduced from the current prices of polycrystalline solar modules.

Hybrid modules are now available which combine conventional crystalline cells with thin film to give the best overall performance. They cost slightly more to produce than crystalline cells but have efficiencies of up to 21%.

The new dye-sensitised solar cells, using principles based on photosynthesis, are not yet at commercial breakthrough level, but expected to be so in three or four years. Their chief advantage is that the photo-sensitive dyes can be printed on many types of surface.

**FIGURE 3.1.2b. Dye-sensitised PV printed onto windows, providing shading from solar glare in South Korea.**

**SOURCE:** Dyesol

## 3.2 **Power output**

The amount of power produced by a solar cell depends on its efficiency and how much light is hitting it. A c-Si cell will produce most power when pointed directly at the overhead summer sun on a bright, clear day.

To enable them to be compared, modules are characterised by 'peak power' (watts-peak or Wp) output. It does not mean that they will always produce this amount of power. Peak power, as measured under internationally agreed Standard Test Conditions (STC), is what is produced when they are exposed to 1 kW per square metre of light at 25 °C in an atmospheric air mass of 1.5.

**EXAMPLE:** *a module that is 10 x 10cm (0.1 $m^2$) and 12% efficient will produce 1.2 W under STC (10% x 12% x 1 kW). Therefore it would be rated at 1.2 Wp. If it was three times as big it would produce three times as much power: 3.6 Wp. If 10 of them were connected together they would produce 10 times as much power (12 Wp), and so on. However, what we're really interested in is the output of the whole system. This is examined later.*

In a southern UK climate, an installation with 1 kWp typically produces 850 kWh of electricity per year (at favourable orientation and shading). Assuming a 16p per kWh feed-in-tariff this equates to an income of £136. A 4 kWp system might cost around £8000. This will take around 15 years to pay off, representing a simple interest rate of 6.8%.

### 3.2.1 Lifetime

Output can tail off over time. After 20 years silicon-based modules may produce 90% of their power, but last another 15 years. The quality of encapsulation can strongly affect longevity, just as it does with OLED computer monitors.

## 3.3 **Environmental impact**

PV modules, when generating power, produce no pollution or greenhouse gases. However, this is not true for their manufacture and that of their system components, their assembly and eventual decommissioning.

Like the rest of the semiconductor industry, many potentially dangerous chemicals are used in the production of silicon solar cells, such as F-gases, including sulphur hexafluoride (SF6), nitrogen trifluoride (NF3) and other climate-affecting and ozone-depleting fluorine compounds. The silicon thin-film industry is in the process of substituting SF6 and NF3 with climate-neutral gases.

The Silicon Valley Toxics Coalition's Solar Scorecard (**http://solarscorecard.com**) tries to keep track of which solar panel producing companies are more environmentally sound than others.

Many panels may contain hazardous waste. In the European Union, the Restriction of Hazardous Substances (RoHS) directive prohibits the use of several toxic chemicals (including cadmium and lead) but the PV industry is currently exempt from this directive. Production facilities based in developed countries with tight environmental legislation will be more likely to have better protection than those companies where production is outsourced to areas of the world with weak legislation and monitoring.

### 3.3.1 Recycling

Silicon, silver, indium, gallium, selenium, tellurium, germanium are all metals used in PV cells. Less than 1% of the last five are currently recycled, yet they are rare and valuable.

Soon, however, PV panels will have to be recycled. Under a new version of the Waste Electrical and Electronic Equipment (WEEE) Directive, from around 2016/17 European member states will have to meet an interim collection target of 45% of all WEEE (including PV system components) placed on the market, rising to 65%, or possibly 85%, of WEEE generated each year around 2020. Collection sites for these panels and the processing system have yet to be worked out.

## 3.3.2 Energy payback

Energy payback is a measure of how long it takes for the modules to generate the energy used to manufacture them. As with cost payback, this depends on where they are installed. Sunnier locations will pay back their energy sooner than cloudy or high latitude ones. For comparison, in central Europe, the following has been found:

**TABLE 3.3.2. Energy payback by cell type**

| Cell type | Energy payback (years) | Percent of total energy that is pollution-free |
|---|---|---|
| c-Si - polycrystalline | 2.2 | 91.2% |
| c-Si - monocrystalline | 1.7 | 93.2% |
| a-Si - ribbon | 1.6 | 93.6% |
| CdTe | 0.7 | 97.2% |
| DSSC | 0.4 | 98.4% |

**SOURCE:** James Durrant, Imperial College, London, UK.

Factoring in the energy payback of the mount and system components would add on average another 15% of time for c-Si and 30% for thin film.

CHAPTER 4

# Applications

**THERE IS AN INCREASING NUMBER** of applications of PV, some of them quite specialised, others niche: all are growing. Each has its appropriate technology solution, depending upon its use and location.

## 4.1 Grid-connected – small-scale

These are now the most common type of application. Typically, they are roof-mounted on buildings but in some cases integrated into the building structure themselves.

Grid connected, small-scale systems are an example of distributed generation, sometimes known as decentralised, embedded or on-site generation. Other examples are: combined heat and power stations, small-scale wind power and small-scale hydroelectric power. By consuming the power close to the point of generation they avoid losses of electrical power due to resistance in cabling over long distances.

Installation requires negotiation with the utility or local electricity grid managers, typically handled by the installer. They will have specifications about the type of inverter, voltage quality, current output, and auto-shut-off.

Electricity supplied by the panels is consumed directly when they are generating sufficient power, or from the grid at other times. Surplus electricity is exported to the grid or supplied to a local grid. Different rates are charged in the UK for that which is sold compared to that which is bought, in accordance with the feed-in tariff rate for the size or type of installation.

## 4.2 **Grid-connected – larger or utility scale**

Otherwise known as solar farms, these arrays of usually ground-mounted panels vary in size up to many megawatts.

The largest in the world is currently a 200 MW installation which was completed in October 2011 in a desert in Golmud, Qinghai Province, NW China. Many other power stations of 100–200 MW are being

**FIGURE 4.2. The UK's largest solar farm**

Solarcentury's 4.99 MWp installation in Wilburton, Cambridgeshire; 19,960 panels generate enough electricity to power more than 1,350 homes.

constructed in China under guaranteed fixed price contracts with the state of 1.15 yuan per kWh, which is very attractive to developers. Canada's 97 MW Sarnia Photovoltaic Power Plant is the largest outside of China, followed by Germany's 'FinowTower' project, expanded in 2011 by 60.4 MWp to 84.7 MWp and therefore currently the largest PV power plant in Europe. Next is Italy's 84.2 MW Montalto di Castro Photovoltaic Power Station and Germany's 80.7 MW Finsterwalde Solar Park.

Tiny in comparison, the UK's largest is Westmill solar farm, located just outside Swindon, a 5MWp community-owned ground-mounted array that is connected to the National Grid under the government's Feed-in Tariff scheme. It generates 4.4 GWh per year. But this will soon be beaten: Lark Energy is currently developing a 30MW (megawatt) solar farm at Wymeswold near Loughborough.

**FIGURE 4.2.1. FinowTower PV power station in Germany.**

### 4.2.1 The advantage of speed

Solar power stations can be erected quickly. The 60.4 MW addition to Solarhybrid's FinowTower were added in just 15 weeks. The first 39 MW of this company's 123 MW Vega plant in Italy was completed by March 2012 in four months. The secret of this speed is good partnerships with other companies. Solarhybrid's partners in this venture include Ja Solar (modules), SMA (inverters), Conecon (assembly of a section) and Enerparc (DC planning), Hilti (mounting systems), Biosar (assembly of a section), BFP (AC planning) and Siemens (transformer station).

## 4.3 **Building Integrated PV (BIPV)**

It used to be said that it was a good idea to fabricate building components that generated electricity; these could be roofs, walls or windows. It was claimed that as the roof/window/wall had to be made anyway, this reduced the overall cost of making and installing the PV element. Experience shows that while sometimes true, this depends upon many factors:

- the correct orientation for the location
- the absence of shading at any time of the day or year
- extra installation costs
- the need to keep the panel cool for optimum performance
- the correct tilt angle for the latitude
- whether building occupants understand and support the installation.

It may instead be more cost-effective to mount efficient PVs at the optimal tilt and orientation in a never-shaded aspect, perhaps on a roof.

**FIGURE 4.3. The solar facade installation by SolarCentury on the Co-operative Group's headquarters, the CIS tower, Manchester, UK.**

## 4.4 Stand-alone systems

These are technically the same as small grid-connected systems, except not connected to the grid. Therefore some sort of storage system, usually batteries, is required. Systems can be DC or AC. If the latter, an inverter is used, but of a different type than for grid-connected systems.

### 4.4.1 Telecoms and weather stations

PV is often chosen to power remote repeater stations (devices that amplify and transmit cellphone signals), other components of mobile networks, and weather monitoring stations because of their reliability and low maintenance. Again, to guarantee a constant supply, hybrid systems are often installed.

## 4.4.2 Road furniture

The same benefits make PV popular for street furniture, such as lights (whether illuminating traffic signs or bus stops), parking meters or traffic lights. Although grid-sourced electricity is available in urban contexts, a PV system can still be attractive to local authorities because it provides an all-in-one turnkey solution without having to dig up the roads to make the connection. It also means that, in the event of grid failure, the traffic system will not be disrupted. Installation comes with a service agreement. Sufficient battery backup is part of the system, with the optional addition of a small wind turbine to charge the battery during sunless periods.

**FIGURE 4.4.2. Solar powered traffic sign in Wales.**

In all cases, bespoke solutions with built-in batteries are available.

## 4.4.3 Gadgets

Many types of appliance now come readily adapted for solar power. Some have integrated solar cells, like radios, watches and lamps. Most of the lamps contain LEDs because of their low power requirements and long life. For items that require more power, such as laptops, televisions and fridges, specially tailored modules are available which supply the appropriate

**FIGURE 4.4.3. Solar powered security system.**

current. Flexible panels can charge personal electrical equipment, for example while hiking.

## 4.5 **Transport**

PVs have many applications across all modes of transport, even trains: in Belgium, a two-mile-long tunnel on the Paris to Amsterdam line built to protect trains from falling trees, has now been covered with 16,000 modules. The electricity produced is equivalent to that needed to power all the trains in Belgium for one day per year.

**FIGURE 4.4.4. Solar powered single-pilot plane.**

For boats, caravans and yachts, combined micro-wind turbine and PV modules systems are available with batteries for storage. That way, power can be supplied both when the wind is blowing and when the sun is shining. Some models of electric car have PVs on them to help power onboard equipment. Solar cars exist which are more for enthusiasts than serious use. But PVs on light aircraft, aerial drones and blimps have more serious potential, in military (drones), leisure and freight or tourist-carrying (blimps) applications.

## 4.6 **PVs in space**

The first applications for PV modules were for powering space satellites. They continue to be used for this purpose, but because of the unique conditions found in space and the stringent reliability requirements, only the most expensive and cutting-edge cells are used. Each mission has its own requirements.

CHAPTER 5

# The Cost of PV Systems

**THE MAJOR COSTS** for solar PV systems are the upfront capital cost and the cost of financing. Operating and maintenance costs are low. Modules account for 30–60% of total system costs (depending on application and technology).

By contrast, fossil fuel generation has comparatively lower upfront costs per kW and high (fuel and maintenance) operating costs. Therefore, cost comparisons between the two, or between PV and other renewable energy technologies, are done on a lifetime average cost basis, otherwise known as the 'levelised cost of energy' (LCOE).

## 5.0.1 The 'levelised cost of energy' (LCOE)

LCOE is used by suppliers of PV electricity to set a unit kWh price for their consumers. The point at which this supplied price matches that of the local price for grid-supplied electricity is called 'grid parity'. This varies across the world, according to the amount of local insolation, the cost of each installation and local energy costs.

Nowhere in the UK is grid parity for PV found, but it has been obtained in other locations (see 5.0.2). As installed costs continue to decline, grid electricity prices continue to escalate and industry experience increases, 'PV will become an increasingly economically advantageous source of electricity over expanding geographical regions', concludes a review of PV's LCOE published in December 2011 (**http://hdl.handle.net/1974/6879**).

The cost of financing is measured by the 'discount rate'. This economists' term is determined by the annual interest rate charged on upfront costs divided by the capital including that interest, averaged over the payback period of the loan. It involves assuming that money 10 years from now is not worth as much as it is today. One consequence of this when applied to the capital funding of energy generation plant is that it makes a fossil fuel plant seem a comparatively better investment, even though it will have major operating expenses in the future while solar PV will not. But if it is recognised that the rate of increase of energy prices can be higher than the discount rate then an appraisal that takes account of the low maintenance costs and slow deterioration of c-Si PV and performs an economic comparison over a 30-year lifetime would be more likely to favour investment in PV.

## 5.0.2 Cost uncertainties

There is huge uncertainty in any LCOE estimate, even for the mature technologies of gas CCGT and coal (see the Mott MacDonald study released by DECC in June 2010, available at **http://www.decc.gov.uk/assets/decc/statistics/projections/71-uk-electricity-generation-costs-update-.pdf**). LCOE estimates are sensitive to eight key factors: the type of technology, the choice of discount rate, the average system price, the financing method, the assumed system lifetime, the assumed future carbon price, the assumed future fuel price and the assumed degree of energy generation degradation over the lifetime.

**FIGURE 5a. An estimate of the levelised cost (£/MWh) of the main electricity generating technologies in 2013.**

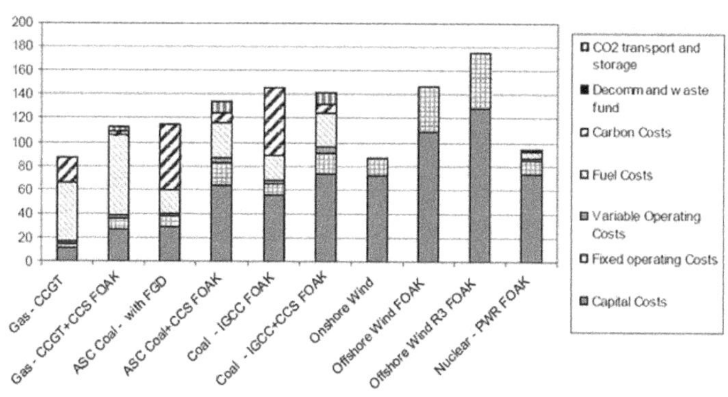


**SOURCE:** Mott MacDonald.

**FIGURE 5b. An estimate of the levelised cost (£/MWh) of the smaller electricity generating technologies in 2009.**

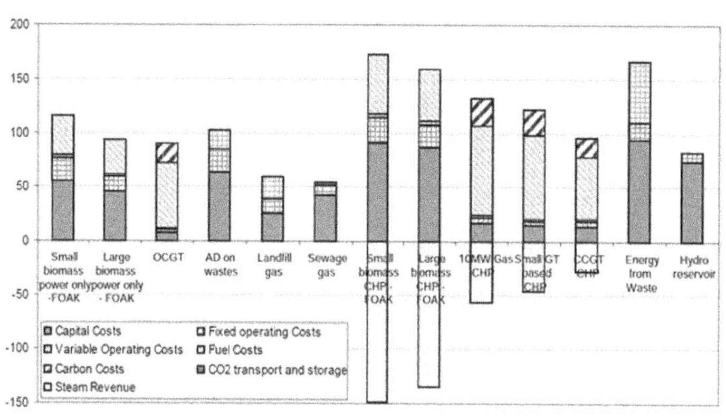


**SOURCE:** Mott MacDonald.

Within these uncertainties, the levelised cost of c-Si PV over 25 years in central Europe is currently estimated at 0.2 euro/kWh (200 euro/MWh), according to researchers at Imperial College, London. Organic PV, currently with an efficiency of 7%, is projected to achieve 10% after three years, and 0.2 euro/kWh after six years, and 0.1 euro/kWh (100 euro/MWh) after 15 years. Their stability will increase to last over 10 years.

For comparison, Figure 5a shows UK government advisors Mott MacDonald's estimate of the projected levelised cost of electricity from different major sources in the UK for plants begun in 2013. Figure 5b shows similar estimates for lower capacity plants begun in 2009. Note: these costs are only those borne by the owner in relation to its operation of the asset. No account is taken of impacts on the wider electricity system, or externalities such as pollution, or fuel supply chain impacts, except those internalised through the price of carbon.

Mott MacDonald acknowledges that any change in the assumptions behind the figures will change the picture dramatically. Furthermore, the overall levelised cost per kWp of a conventional c-Si installation depends on its size. Any reader looking for certainties will fail. Energy is a business that can only seek risk reduction.

### 5.0.3 Grid parity

Solar PV will most quickly become competitive in places where there is higher solar insolation, electricity rates are high or where utilities have inordinate monthly charges. This means that in Europe, Italy, where electricity costs are much higher than in France, will achieve grid parity first, as seen in Figure 5c.

**FIGURE 5c. Projected grid parity for solar PV in Europe in 2018.**

This shows three major market segments: residential (orange), small commercial (yellow) and utility level (blue). The circle size indicates estimated country's market volume in 2018. The assumed learning rate is 15% (green line) and 20% (red line).

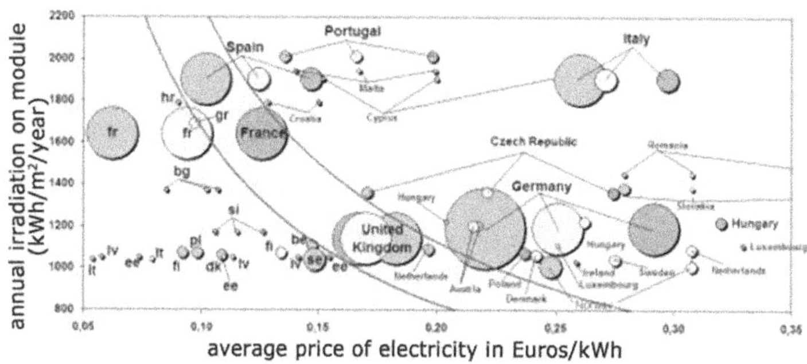


**SOURCE:** Dr Chiara Candelise, Imperial College, London, UK.

Within eight years, at the current phenomenal rate of cost reduction, the forecast is that consumer prices will be between €0.08 and 0.18/kWh in 2020 (depending on the application), matching the price of conventional grid electricity in many areas of Europe. As the EPIA says in 2011's market survey: 'the price of PV modules has decreased by over 20% every time the cumulative sold volume of PV modules has doubled'. Whether this includes the UK remains to be seen.

Thin film technologies are expected to reach grid parity earlier than c-Si. Key drivers for c-Si cost reductions are: the reduction in the amount of silicon and other materials used in the final module; new and improved silicon feedstock and wafer manufacturing technologies; an increase in cell and module efficiency; and faster production processes.

Key drivers for thin film cost reductions are: increases in efficiency (which also reduces area related BOS costs); lower and more efficient materials usage; and increased demand/production lowering unit costs.

## 5.1 **Capital costs**

As we have seen, and as is demonstrated in Figure 5.1, module costs are falling, mainly due to increased capacity of producers in the Far East.

**FIGURE 5.1. As PV module production capacity has increased, the price has fallen.**

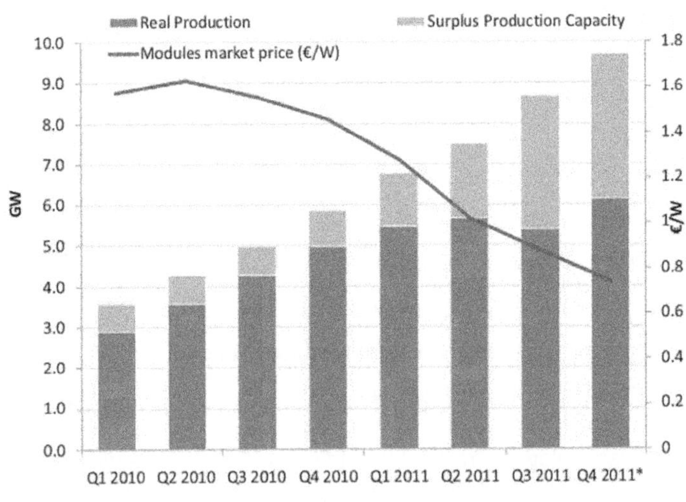


**SOURCE:** IHS.

Capital costs for medium- to large-scale projects are primarily in construction (an average of 92% of total), with pre-development costs averaging 6% and grid connection costs 2% (DECC, 2011). This breakdown is not applicable for micro-scale solar.

The bottom line is that the larger the system, the cheaper per MW output. 2011 UK Government estimates for future costs, shown in the tables below, reveal that the future capital costs of a system above 50 kW in size are about two-thirds per kWp that of one whose overall size is below 50 kWp, while the operating costs are also lower.

**NOTE:** Low, Median and High refer to scenarios which imagine the maximum amount of capacity that could be installed per year between now and 2030 in the UK given current constraints (Low), if some of the constraints are relaxed (Median), and if additional constraints are relaxed (High):

**TABLE 5.1a. System <50kW – capital costs projections (real), £000s/MW**

| | 2010 | 2015 | 2020 | 2025 | 2030 |
|---|---|---|---|---|---|
| High | 5080 | 4287 | 3640 | 3254 | 3016 |
| Median | 3339 | 2647 | 2115 | 1814 | 1634 |
| Low | 2732 | 2027 | 1520 | 1248 | 1091 |

**TABLE 5.1b. Array <50kW – operating costs projections (real), £000s/MW**

| | 2010 | 2015 | 2020 | 2025 | 2030 |
|---|---|---|---|---|---|
| High | 71 | 71 | 71 | 71 | 71 |

| | | | | | |
|---|---|---|---|---|---|
| Median | 25 | 25 | 25 | 25 | 25 |
| Low | 17 | 17 | 17 | 17 | 1 |

**TABLE 5.1c. Array >50kW – capital costs projections (real), £000s/MW**

| | 2010 | 2015 | 2020 | 2025 | 2030 |
|---|---|---|---|---|---|
| High | 3736 | 3153 | 2677 | 2393 | 2218 |
| Median | 2710 | 2148 | 1717 | 1472 | 1260 |
| Low | 1873 | 1390 | 1042 | 855 | 748 |

**TABLE 5.1d. Array >50kW – operating costs projections (real), £000s/MW**

| | 2010 | 2015 | 2020 | 2025 | 2030 |
|---|---|---|---|---|---|
| High | 27 | 27 | 27 | 27 | 27 |
| Median | 21 | 21 | 21 | 21 | 21 |
| Low | 16 | 16 | 16 | 16 | 16 |

**SOURCE:** Department of Energy and Climate Change, 'Review of the Generation Costs and Deployment Potential of Renewable Electricity Technologies in the UK', June 2011.

This report notes that the key cost drivers are considered to be module costs, labour rates and exchange rates. The most significant decline in costs is forecast to happen between 2010 and 2020 as global deployment rapidly scales up. During this period, costs are anticipated to fall by 37%, compared with an overall 51% decline in capital expenditure expected by 2030.

## 5.1.1 Operating costs

Operating costs for solar PV claimed in the above study include: labour, maintenance, inverter replacement, land rental, insurance and grid charges where applicable. For micro-scale solar projects (<50 kW), costs claimed by developers range from £17,000/MW/year to £71,000/MW/year, with a median of £25,000/MW/year, reflecting the uncertainty inherent in a developing domestic market. The median operating cost quoted by developers for the study was £58,000/MW/year.

Larger solar PV projects (>50 kW) O&M costs can expect annual fees between £16,000/MW/year and £27,000/MW/year, with a median of £21,000/MW/year, depending on the project, with fees being lower per MW for large-scale ground-mounted solar installations and highest for geographically dispersed roof-mounted installations. As there are few operational large-scale solar projects in the UK, these costs are based on European averages and liable to change.

## 5.1.2 Cost comparisons

In the UK, solar power will only ever make up a tiny proportion of overall generation because the country does not benefit from high insolation, and it is not currently cost-effective compared to other actions that save carbon. The Department of Energy and Climate Change's 'Review of the Generation Costs and Deployment Potential of Renewable Electricity Technologies in the UK' (June 2011) concludes nevertheless that 'This is a technology with very significant deployment potential of 16.6GW by 2030 (medium forecast), but with very high capex', but capital expenditure costs per MW go down as the size of installations increases, as Figure 5.1.2 shows.

**FIGURE 5.1.2. December 2011 chart from DECC's Impact Assessment of Renewable Obligation banding, comparing PV with other technologies. By 2015 PV comes out favourably compared to many other technologies.**

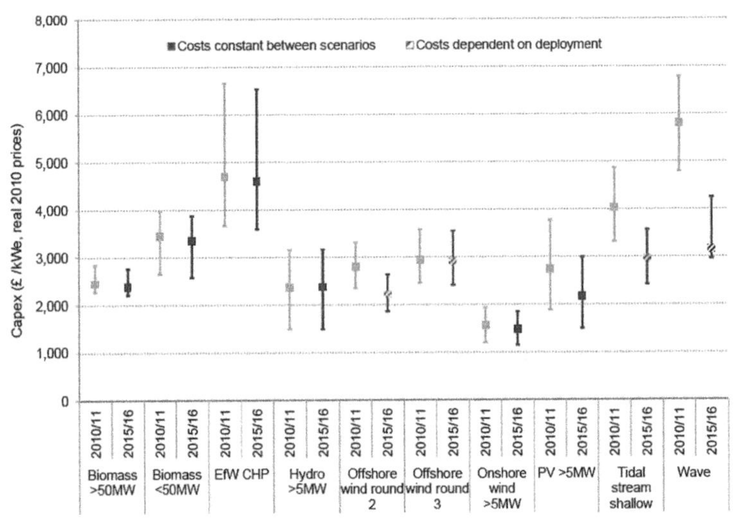


Note: Where costs are dependent on deployment, the cost for the Extra Support for Marine option is shown.

Source: DECC, Arup, Ernst and Young

Comparing levelised costs of energy, solar PV was in 2011 the most expensive form of renewable energy per megawatt-hour in the UK at a smaller scale of 50 kW–200 kW: its levelised cost was three times that of onshore wind, and 11 times that of energy-from-waste plants. See Tables 5.1.2a and 5.1.2b.

**TABLE 5.1.2a. Levelised cost of electricity given a 10% discount rate, 2011 project start, at projected Engineering Procurement Contract prices, First-of-a-Kind/Nth-of-a-Kind mix:**

| Technology | Levelised cost £/MWh |
|---|---|
| Energy from waste | -30 |
| Landfill gas | 44 |
| Hydro 5–16 MW | 73 |
| Cofiring conventional | 74 |
| Sewage gas | 83 |
| Onshore wind >5 MW | 91 |
| Anaerobic digestion <5 MW | 105 |
| Biomass 5–50 MW | 8 |
| Offshore wind > 100 MW | 31 |
| Hydro <5 MW | 131 |
| Biomass CHP | 134 |
| Bioliquids, biodiesel | 306 |
| Solar >50 kW | 314 |

**TABLE 5.1.2b. Levelised cost of electricity given a 10% discount rate, 2017 project start, at projected Engineering Procurement Contract prices, all Nth-of-a-Kind**

| Technology | Levelised cost £/MWh |
|---|---|
| Onshore wind > 5 MW | 88 |
| Offshore wind > 100 MW | 117 |

| | |
|---|---|
| Biomass > 50 MW | 122 |
| Biomass 5–50 MW | 126 |
| Offshore wind, Round 3 | 130 |
| Solar | 241 |

**SOURCE:** Department of Energy and Climate Change, 'Review of the Generation Costs and Deployment Potential of Renewable Electricity Technologies in the UK', June 2011.

If the policy goal is to reduce carbon emissions in the domestic sector, where most PV is installed in the UK, then PV is also one of the most expensive means of doing this, as measured in pounds sterling spent per tonne of carbon dioxide saved. Table 5.1.2c compares the cost in 2008 (latest figures) of investing in PV not just with other renewable energies but with energy efficiency measures as well. In the list of measures in this table, PV fares poorly in terms of cost-effectiveness. Wood pellet boilers and several forms of insulation are all around seven times cheaper and ground-source heat pumps are five times cheaper.

**TABLE 5.1.2c. Kilograms of carbon dioxide saved per pound spent per measure**

| Measure | Kilograms of $CO_2$ saved per pound |
|---|---|
| Converting an existing community heat to CHP | 88 |
| SWIa external (semi-detached house) | 25 |
| Wood pellet boilers (primary) | 24 |
| Loft insulation (professional, from zero) | 21 |
| Ground source heat pumps | 14 |

| | |
|---|---|
| Replacing old boiler (65% efficient by 88.3%) | 13 |
| Glazing upgrade E–C rated | 8 |
| Heating controls - upgrade with new heat system | 6 |
| Solar water heater (4 m) | 4 |
| Photovoltaic panels (2.5 kWp) | 3 |
| Micro wind (1 kWp, 1% LF) | 0 |

[a] SWI = solid wall insulation.

**SOURCE:** 'Explanatory Memorandum to the Electricity and Gas (Carbon Emissions Reduction) Order 2008'.

A similar conclusion is obtained when comparing the cost in 2008 of saving one tonne of carbon dioxide by employing different methods in the domestic sector. Here, district heating and power schemes using woodchip fuel are 11 times cheaper than PV, and ground-source heat pumps are eight times cheaper. However, since 2008, costs of PV have more than halved and continue to fall.

It is logical then to ask: if PV is expensive, comparatively speaking, then why use it at all? There are two answers to this: location/application and policy aims.

### 5.1.2.2 Location drivers

Many renewable energy technologies are site-specific. They either depend on the local geography, like an appropriately windy site (wind power), a nearby water course (hydro power) or marine currents (marine current turbines). Or they require a reliable stream of fuel supply such as biomass (wood pellet/chip stoves or boilers), burnable waste (energy-from-waste) or organic matter (anaerobic digestion). Other technologies, such as heat pumps or bioliquids, don't suffer from this disadvantage.

Silicon or thin film PV is site-specific in the sense of its need for unshaded, appropriately angled locations. Such locations may not have any other renewable resource available; for example in urban areas the wind is generally too turbulent to be useful for wind turbines, but there are plenty of empty roof spaces. Here, its use may be welcomed if one is not worried solely about the return on investment but wants to feel better about using a clean source of electricity. There may also be other incentives or legal requirements to cut carbon use.

#### 5.1.2.3 Policy drivers

All new energy technologies need support at their start in order to bring them to a stage at which they can survive in the market unaided. In fact, the petrochemical industry is heavily subsidised too. In 2008–2010, fossil fuels were being subsidised globally by almost 13 times the level of renewable energy sources such as wind, solar power and biofuels, to the tune of $557 billion compared with $43–46 billion for renewables, according to Bloomberg New Energy Finance (**http://www.bloomberg.com/news/2010-07-29/fossil-fuel-subsidies-are-12-times-support-for-renewables-study-shows.html**). Rather than going down, fossil fuel subsidies are increasing. The IEA expects them to reach $660 billion, or 0.7% of global GDP by 2020. In this context, support for PV is essential to give it a chance of becoming a big player and help fight climate change.

## 5.2 **Feed-in Tariffs**

The difference between the wholesale price of energy and the subsidy rate is called the premium price. Feed-in Tariffs are currently the UK Government's main policy for supporting PV. The stated policy aims at the start in 2010 were twofold: to create more demand for modules in order to ramp up production and bring their unit price down; and to increase public awareness of renewable energy. In 2011 both these aims succeeded beyond expectations. As a result, the tariff rate has been reduced from 43p to 16p per kWh for common domestic installations. The costs of modules have also almost halved.

The current rates are:

**TABLE 5.2. Current UK Feed-in Tariff rates**

| Band (kW) | Standard generation tariff (p/kWh) | Multi-installation tariff (p/kWh) | Lower tariff (if energy efficiency requirement not met) (p/kWh) |
|---|---|---|---|
| <4kW (new build and retrofit) | 16.0 | 14.4 | 7.1 |
| >4-10kW | 14.5 | 13.05 | 7.1 |
| >10-50kW | 13.5 | 12.15 | 7.1 |
| stand-alone | 7.1 | n/a | n/a |

The multi-installation tariffs are set at 80% of the full rate, and are for developers like local authorities or housing associations. To be eligible for the full tariffs, a building must have an Energy Performance Certificate rating of level D or above, regardless of whether it is a domestic or non-domestic building, otherwise it will be given the lower tariff rate. This is to ensure that buildings meet minimum energy efficiency standards.

Properly insulating buildings saves five times more greenhouse gas emissions per pound spent than the 21p Feed-in Tariff for solar electricity.

The rationale for making these reductions is sound, but the speed at which they have been made has dismayed the solar industry, forcing business plans to be torn up, and having a poor effect on investment decisions.

Utility-scale solar installations currently receive support under the Renewables Obligation (RO) worth 2 ROCs (Renewables Obligation Certificates per MWh). This makes non-domestic solar competitive with other renewable technologies supported under the RO. Around 500 MW of solar was expected this year under the RO. The UK Government is currently consulting over whether to withdraw this support from April 2013 for schemes under 5 MW. These would have to look to the Feed-in Tariff scheme for support, which has a much lower budget. The Solar Trade Association is concerned that the effect of this would be to greatly limit support for this scale of deployment of the technology, whose developers are quite different from those at the domestic scale, often farmers and other landowners looking to diversify their income.

In other countries such as Denmark and Germany there is much more community ownership of renewable energy projects and there are moves to repeat that success here. The UK's Minister for Energy and Climate Change, Greg Barker, has publically committed himself to the belief that community-backed schemes that help put solar modules on offices, homes, churches and public buildings are key to the uptake of renewable energy in the UK. The Department of Energy and Climate Change (DECC) has said that community-owned, social housing and the commercial rent-a-roof sectors' projects of 50kW or less (DNC) will be eligible for a generation tariff equal to 90% of the standard tariff.

The Co-operative Enterprise Hub [**http://www.co-operative.coop/enterprisehub/**], run by The Co-operative, has committed £6m between 2012 and 2014 to enable it to deliver free advice and guidance to create and grow sustainable member-owned enterprises across the UK. The Co-operative also runs the Co-operative Energy Challenge [**http://www.co-operative.coop/join-the-revolution/our-plan/Environment/Community-Energy-Challenge/**], which aims to provide financial backing and support, to a select group of communities across the UK to help them develop significant renewable energy projects.

The Westmill community scheme described above is designed to raise £4 million equity with potential investors able to buy-in to the project with a minimum investment of £250 and a maximum of £20,000. Many community solar initiatives are co-operatively owned, and therefore registered as an Industrial & Provident Society with the Financial Services Authority in England and Wales under the Industrial Provident and Provident Societies Act 1965. This enables them to issue shares.

**FIGURE 5.2a. Subsidies for PV have suffered much more volatility than for wind power over the last two years.**

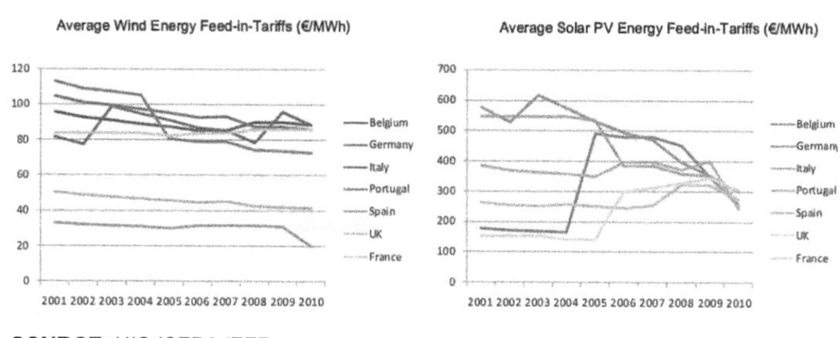


**SOURCE:** HIS/CERA/EER.

For example, The Solar Coop in Maidenhead has been instrumental in supporting 20 installations.

Feed-in tariff rates have been reduced by similar amounts in many places, and the cause is more than just the fall in module prices. Subsidies for PV have suffered much more volatility than for wind power over the last two years, partly due to installed PV capacity rising faster than predicted, causing increased balancing costs for grid regulators to maintain the stability of the grid. In the recession, governments are also worried about household end-user energy costs and fuel poverty. While high oil and gas prices is the real reason for high energy bills, governments have no control over this but can control subsidies to renewables.

Subsidies to PV come under greater pressure than other renewables because the ratio of subsidy to power output is higher for PV than any

**FIGURE 5.2b. Projected subsidies till 2020 for all European countries.**

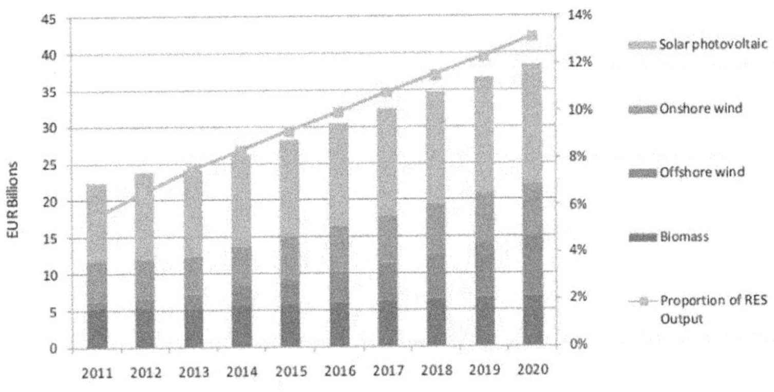


**SOURCE:** HIS/CERA/EER.

other renewables. In Europe, of the €22bn spent subsidising renewables last year, half went to solar PV, but its output constituted less than 6% of renewables' total output over Europe as a whole.

This trend is projected to remain largely unchanged. Subsidies up to 2020 for all renewables are expected to rise to €35bn, and solar PV will continue to take about half of this, but total output is not expected to exceed 14% of all renewables' generation. In other words, solar PV provides poor value for money at present, but prices are coming down

**FIGURE 5.2c. The UK is perceived as more at risk from policy changes but less risky from interest rate changes than most other European countries.**

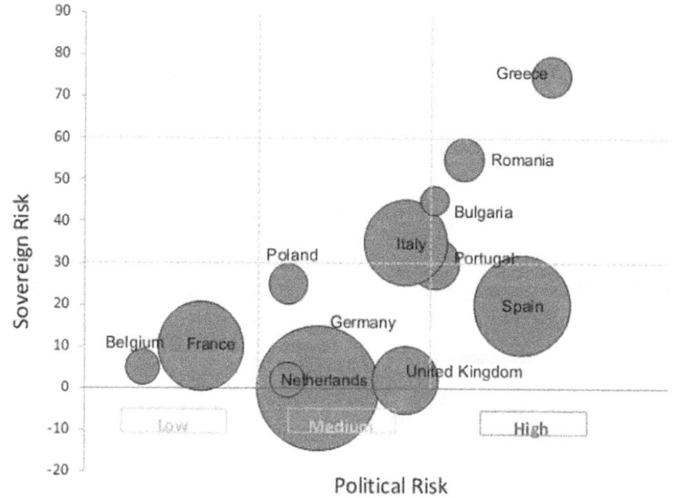


**SOURCE:** HIS.

and the industry will mature. Moreover, the price is not the only issue as described above.

Energy consultancy IHS calculates that there is a high political risk but a low sovereign risk to renewable energy subsidies in the UK; in fact it is Southern Europe that presents the most uncertainty to investors over unplanned cuts to subsidy levels. Belgium and France provide the most secure outlook.

### 5.2.1 Capacity trigger

In the near term, the UK Government is considering a 'capacity trigger' system to reduce the amount of subsidy solar photovoltaic systems are given, and spread it over more installations. This will mean that the amount of the tariff is reduced depending on the generation capacity of each installation, as in Germany where the tariff is calculated by adding an overall reduction of the tariff (9%) to an amount relative to what is installed. For example, an installation above 3500 MW of new PV capacity would entail a 12% reduction, while 5200 MW would entail a 15% reduction. At the same time, feed-in tariffs for solar modules are being reduced by 15% in 2012. Under Germany's new Renewable Energy Action Plan, solar PV will generate 41 TWh of electricity per year for as much as 7% of 2020 consumption.

The UK's Renewable Energy Association has indicated that a capacity trigger 'could be helpful to us' as 'it would mean that if the price of panels doesn't come down as anticipated, the model would account for that'.

A drawback of capacity triggers is that, per installed unit of capacity, the cost of installing many separate systems is greater than that of installing a smaller number of higher systems. Therefore the cash available will support less, not more, generation capacity.

## 5.3 **State of the industry**

Globally, the PV industry generated US$82 billion in revenues in 2010 and energy demand is projected to increase by 49% over the next 30 years.

### 5.3.1 Long-term outlook

The long-term outlook is good. The IEA's 2011 World Energy Outlook forecasts that photovoltaics will generate 4.5% of the world's total electricity supply by 2035, or 740 terawatt-hours (TWh), an increase from 20 TWh in 2009, implying a growth rate of 15% per year. Total renewable generation will account for 44% of the electricity generated from sources installed post-2009. Growth in the UK is likely to be on a par with this.

IEA figures also forecast that while 92 GW of photovoltaic capacity will be retired between the 2009 to 2035 report timeframe, 553 GW will be installed, resulting in a net increase of 461 GW.

But consultant Veit Robert Otto from EuPD Research believes this figure is too low and that photovoltaics are on a track to take a far greater share of the electricity market by increasing output through technical innovation. 'I think we are on a good track. We see on these initiatives like Desertec, they have transferred the technology from CSP to photovoltaics and this is another indicator of the competitiveness of photovoltaics,' he says.

## 5.3.2 Desertec

The Desertec project, a German-based group with commercial backers including Siemens and Deutsche Bank, aims to power Europe and North Africa using renewable energy plants across the Middle East and the North African coast, in particular using concentrating solar thermal (CSP) power, photovoltaics and wind energy. CSP has been getting cheaper, but not as quickly as PV power, causing their plans to change emphasis.

A study by the German Aerospace Centre (DLR) concludes that less than 1% of suitable land would be needed to cover the current electricity consumption of the region, as well as Europe. Desertec has estimated

**FIGURE 5.3.2. The outline plan for the Desertec project which is actively seeking investment.**

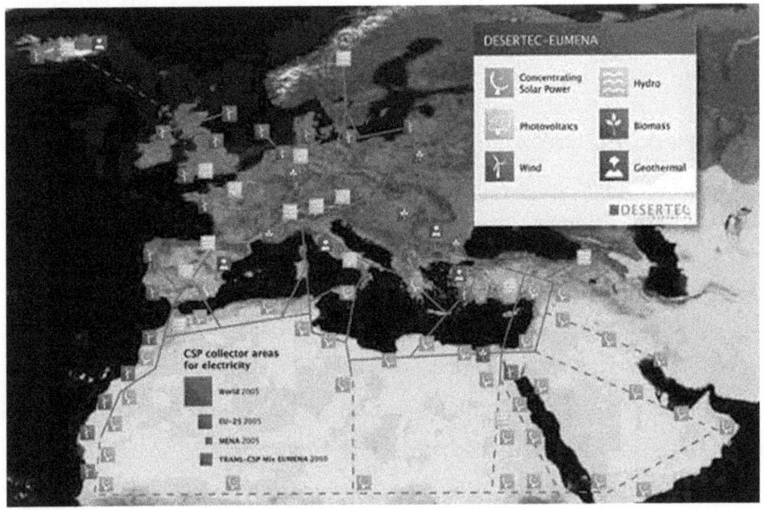

the cost of achieving this by 2050 at $400 bn. Much of this will be investment in the grid connection infrastructure. Others believe that it would be cheaper to generate the same amount of power closer to home. The Desertec industrial initiative sees the priorities as follows: first, avoid wasting energy (energy efficiency); second, encourage decentralised power generation where reasonable and feasible; finally, promote the centralised generation of renewable energy in places with a wealth of natural resources such as solar, wind and hydro power, thus offering the most suitable conditions.

Morocco, which has its own ambitious solar programme, will be the site of its first 'reference' pilot project, with electricity projected to flow through undersea cables from Morocco to Spain from 2014. It has a target of producing 42% of its electricity from renewable sources by 2020.

## 5.3.3 Short-term outlook

In the short term, however, the PV industry is in turmoil, partly as a result of its success. The 2012–2013 forecast is for a period of 'consolidation', with mergers and acquisitions seeing weaker solar production companies being swallowed up. The overall market in Europe is projected to be down 72%, with the ground-mounted (solar farm) segment the hardest hit and the residential sector the least affected (down 41%).

Nevertheless, the overall outlook is positive, since PV is an essential weapon against climate change and for rural electrification. A useful way of thinking about the market situation is to see it as comparable with the dot.com bubble burst of 1999.

## 5.3.4 Outlook for solar farms

The average time taken to permit and build a 1 GW nuclear power plant is 13 years, whereas the corresponding time for a ground-mounted 1 GW solar farm is one year. This, and the exorbitant front-loaded construction costs, helps to explain why nuclear power is in the doldrums compared to solar power, and why its lobbyists are working hard to persuade policy-makers to support it.

The International Energy Agency's 2011 report 'Solar Energy Perspectives' said the following: 'The rapid cost reductions driven by [increased] deployment have confirmed earlier expectations related to the learning rate of PV. They have also increased confidence that sustained deployment will reduce costs further – if policies and incentives are adjusted to cost reductions, but not discontinued.'

For solar farms, the cost of modules at <€1/Wp is less than 30% of system cost. The rest, known as the 'balance of station' (BOS) costs, although reducing in price, do so at a slower rate. They include the cost of financing, the structures, the power-conditioning equipment and the land. However, there is talk in the UK among developers of solar farms up to even 40 MW in size being built in the south of England. REC Solar, Canadian Solar, Q-Cells and many others are all of the opinion that if the component prices decline as expected, and energy bills continue to rise, opportunities for ground-mounted solar to become cost-effective will increase. REC Solar is hoping to double its capacity in the UK in 2012–2013 by installing approximately 60 MW, a large proportion of which will be ground-mounted. Taking all developers' plans together, 1 GW of solar power may well be installed in the UK in the next 18 months.

## 5.4 **Advice to installers**

For a solar business considering offering installation packages to small clients who cannot afford the upfront capital costs: offering zero interest financing remains the most attractive and more reliable incentive than a feed-in tariff or a tax credit.

CHAPTER 6

# Planning a Solar Installation

**THIS SECTION IS FOR THOSE** considering a solar installation.

## 6.1 System components

A typical system includes: the modules, a grid-inverter, which is a different type of inverter to that used in stand-alone (non-grid-connected) systems, controllers, disconnects, meters and fuses. A system that is also required to supply power when the grid is down would include a battery-connected inverter and relay controller.

## 6.2 Calculating output

The annual energy that can be produced by a given array of modules depends upon several factors, most especially the type of module and the location. It is calculated using the annual energy output per peak kilowatt of rated power supplied by the manufacturer of the module and the insolation data for the location. Sources for this are given at the end. This will give the average amount of energy available per square metre for that location - in $kWh/m^2$ - for the whole year.

To calculate the estimated output per year per PV module, use the following formula:

**MODULE RATED OUTPUT (PEAK WATTS)
x PEAK SUNSHINE HOURS (PER YEAR)
x 0.75 (PERFORMANCE RATIO) =
ENERGY GENERATED (KILOWATT-HOURS/YEAR)**

The performance ratio is to cover higher module operating temperatures, system losses and other factors. In a properly installed grid-connected system it is usually between 0.7 and 0.8.

For example, 1500 W x 900 peak hours per year x 0.75 = 1,012,5005 watt-hours (Wh) or 1102 kilowatt-hours (kWh) (approximately)

Other factors to be taken into account include:

- the orientation (azimuth) (–90° is east, 0° is equator-facing and 90° is west);
- the tilt angle of the modules from the horizontal plane, for a fixed (non-tracking) mounting;
- the energy conversion efficiency of the modules;
- the extent to which their efficiency is affected by temperature;
- factors to do with the clarity of the atmosphere and the path of the sun;
- whether there are any obstructions which might cause shade on the modules at any time of the day and year;
- system efficiencies, such as those of the chosen inverter and the wiring;

- the type of mounting structure - fixed or tracking (tracking increases output).

The larger the installation, the more detailed these calculations need to be. *The Solar Energy Pocket Reference Book* (Christopher L. Martin and D. Yogi Goswami, Earthscan, 2005) published by The International Solar Energy Society is a handy source of formulae.

### 6.2.1 UK SAP guidelines

The following very rough way of calculating output is taken from the Standard Assessment Procedure (SAP), the UK Government's tool for assessing the energy and environmental performance of dwellings in the UK.

First, establish the installed peak power of the PV unit (kWp). The electricity produced by the PV module in kWh/year is then:

**0.8 x kWp x S x ZPV (M1)**

where S is the annual solar radiation from Table 6.2.1a (depending on orientation and tilt) and ZPV is the overshading factor from Table 6.2.1b.

If there are two PV modules, for example, at different tilt or orientation, apply equation (M1) to each and add the annual electricity generation.

**TABLE 6.2.1a. Annual solar radiation, kWh/m**

| Tilt of Collector | Orientation of collector | | | | |
|---|---|---|---|---|---|
| | South | SE/SW | E/W | NE/NW | North |
| Horizontal | | | 961 | | |
| 30 ° | 1073 | 1027 | 913 | 785 | 730 |
| 45 ° | 1054 | 997 | 854 | 686 | 640 |
| 60 ° | 989 | 927 | 776 | 597 | 500 |
| Vertical | 746 | 705 | 582 | 440 | 371 |

**TABLE 6.2.1b. Overshading factor**

| Overshading | % of sky blocked by obstacles | Overshading factor |
|---|---|---|
| Heavy | > 80% | 0.5 |
| Significant | > 60–80% | 0.65 |
| Modest | 20–60% | 0.8 |
| None or very little | < 20% | 1.0 |

The figures are a rough guide only since, as can be seen from Figure 2.2.2a, the amount of insolation in the UK varies considerably. The table does indicate the influence of tilt angle on output.

A new change in SAP2012 also specifies that where photovoltaic (PV) panels on blocks of flats are connected to the landlord's supply rather than to the individual flat, the Energy Performance Certificate for that flat will include the relevant share of the carbon savings from the PV panels, but no cost savings will be attributed to the flat. (**SOURCE:** SAP, 2012 technical paper STP11/SR01.)

## 6.3 General design advice

The following advice comes from experience of many systems:

- Keep it simple: increased complexity reduces reliability and increases costs, especially for maintenance.
- Appreciate that the system may not be available 100% of the time. Achieving this makes the system more expensive: be realistic.
- For stand-alone systems, be realistic when estimating loads: including a large safety factor can increase costs substantially.
- Repeatedly check weather data: errors in estimating the solar resource can cause disappointment.
- Different hardware with different characteristics have different costs. They may also be less compatible. Investigate thoroughly all options before deciding on the optimum combination.
- Ensure the system is installed carefully: each connection must be made to last 30 years, because it can if installed properly. Deploy the correct tools and techniques. System reliability is no higher than the weakest connection.
- Electricity is dangerous: be rigorous about safety precautions during installation and in operation. Comply with local and national building and electrical codes.

Plan for periodic maintenance: PV systems have a good reputation for unsupervised operation but all require some degree of monitoring and care.

Calculate the lifecycle costs to compare PV systems with alternatives. This reflects the complete lifetime cost of owning and operating any energy system.

## 6.4 **Where should it go?**

The optimum position is unshaded at any time of the year and faces the equator, that is, its azimuth is 0°. The tilt angle relative to the horizontal should be appropriate to the latitude. There is no easy formula for this; ask the correct angle of each potential supplier, or consult the *Solar Energy Pocket Reference Book*.

The modules should:

- not be shaded, especially between the peak sun hours of 10am to 3pm
- have sufficient space to mount the required number, where there is no risk from damage or overheating, since efficiency tails off significantly at higher module temperatures
- be accessible for regular cleaning
- have a roof or mount able to withstand their weight, wind forces, plus the weight of any snowfall.

## 6.5 **Costs**

Modules account for on average 50% of total system costs (depending on application and technology). To reduce costs (£/kWh) it is necessary to:

- reduce the Balance of System costs (system components and

installation costs);

- increase the energy yields, stability and lifetime of the system;
- increase the inverter lifetime and reliability of system components.

The cost (or value) of solar electricity is much higher in the winter, when light is scarce, than in the summer. For example, 7.5 hours of sunlight might provide the same power, 20 Ah, in the summer, as 240 hours in the winter. In a grid-connected system the shortfall of loss of electricity independence is made up by purchasing grid power. In a stand-alone system it is made up by storage or back-up power. The cost of this needs to be factored in to any decisions.

## 6.5.1 Estimating cost saving

The cost saving associated with the generated electricity depends on whether it is used directly within the dwelling or whether it is exported. Electricity used directly within the dwelling is valued at the unit cost for purchased electricity (standard tariff, or the high/low rate proportions given in Table 5.1a **[AQ]** in the case of an off-peak tariff). Electricity exported is valued at the price for electricity sold to the grid. The effective price depends on a factor b, which is in the range 0.0 to 1.0 and is defined as the proportion of the generated electricity that is used directly within the building. Its value depends on the coincidence of electricity generation and electricity use. The fuel price to calculate the cost benefit is then:

**B x NORMAL ELECTRICITY PRICE + (1 - B) x EXPORTED ELECTRICITY PRICE.**

### 6.5.2 Estimating $CO_2$ emissions saved

To calculate the $CO_2$ emissions saved, the emissions factor is that for the grid-displaced electricity, which will depend on the location of the installation and the make-up of the grid-supplied electricity. In the UK, this is generally taken to be 0.517 kg $CO_2$ per kWh, although in practice it varies. The same factor applies to all electricity generated, whether used within the building or exported. You would multiply the total number of kWh generated by this factor to determine the number of kilograms of carbon dioxide saved.

## 6.6 **Inverters**

If the property is connected to the mains, then it is almost always beneficial to connect the system to the mains as well. This will inevitably entail the use of an inverter. Inverters convert the direct electrical current from the modules into the alternating current used by the mains. The inverter will come in the circuit before all of the loads.

Inverters vary according to the voltage and current they can accept and the frequency and power of the output. The quality of output (waveform) also varies with cost: sensitive electronic equipment needs a smoother supply - a sine wave, which is like normal mains power - than lighting, TVs and power tools, which can accommodate a square wave output. The latter are generally cheaper than the former. Inverters, like all transformers, consume energy, and this wasted energy can be a high proportion of what is generated.

Monitoring equipment is usually installed to record the amount of sunshine and how much power has been generated.

## 6.7 Maintenance

Solar systems require little maintenance. Modules should be inspected and cleaned regularly, particularly if there is snow or a lot of dust in the area. The power output, the module and battery temperature, and perhaps inverter performance should be monitored and recorded, which can be done automatically; this enables understanding of the system and the ability to spot potential problems before they can do harm, or issues which are causing the system to perform inefficiently.

## 6.8 Sourcing and talking to suppliers

Whatever system is to be installed, when sourcing suppliers, it is vital to bear the following points in mind:

1. Always choose a supplier who has been in business for some time and obtain independent testimonials from previous clients.
2. Obtain at least three quotes from different suppliers.
3. Make sure the installers are members of the REAL Assurance Scheme (in the UK) (http://www.realassurance.org.uk), whose members agree to abide by their Consumer Code.
4. Using the information in this publication, prepare a list of questions to ask of each potential supplier.
5. Always be sure to compare like with like; the same number of components with comparable specifications.
6. Make sure they explain how they have calculated the size of the system to be appropriate for your needs, supply clear information

and operating instructions, provide an estimate of how much electricity will be generated by the proposed system, and what this is as a proportion of your annual use.

7. Ask for written guarantees and to see their insurance criteria.

Remember that the most efficient modules or systems are often more expensive, so do not compare on price alone, but on overall cost/benefit over the lifetime of the system (25 years or more).

## 6.9 Insurance

Damage to system hardware should be covered by existing buildings insurance but the policy conditions should be checked. Coverage for electricity not generated due to an insurable event would be expensive but is feasible.

## 6.10 Warranties

Manufacturers offer performance warranties on modules which typically last up to 25 years. Inverter warranties last up to five years. Installers may also offer their own warranties.

## 6.11 Generation data

In order to illustrate the impact of seasonal variation in the UK, here is the actual data for kWh generation of a 58 $m^2$ silicon photovoltaic system in Sheffield, England in 2011.

**TABLE 6.11. kWh generation of a 58 $m^2$ silicon photovoltaic system in Sheffield, England in 2011**

| Month | kWh |
|---|---|
| January | 114.19 |
| February | 183.51 |
| March | 547.29 |
| April | 876.11 |
| May | 1,029.78 |
| June | 1,054.53 |
| July | 993.13 |
| August | 839.48 |
| September | 653.48 |
| October | 374.10 |
| November | 135.23 |
| December | 73.23 |
| TOTAL | 6874.06 |

At the time of installation it was expected to generate around 6500 kWh per annum, so in this year has exceeded it.

CHAPTER 7

# Investing in PV

**THIS SECTION LOOKS AT THE INVESTMENT** case for solar photovoltaics, and the support and advice available in the UK.

Drivers to incentivise the implementation of photovoltaic power include the continued escalation of fossil fuel and energy prices, the urgency of climate change, legislative drivers, European finance availability, the continued lowering price of modules, other investment budgets, drivers arising from the 2012 Earth Summit, and even requirements from some insurance policy providers that their clients take steps to mitigate and adapt to climate change.

Disincentives include the continuing low price of carbon, which for large investors can be a factor if finance is to come from offsetting funds.

Ever-reducing module prices means that installations throughout the world are continuing apace. In 2011, solar power investments were responsible for a massive 49% share of global renewable power investments. This amounts to an estimated €60 billion (US$79 billion) (REN21, 2011, Renewables 2011 Global Status Report). This compares with wind power's 34% share. Venture financing and private equity investments were also high for solar technology developments. With solar power prices approaching grid-parity, asset financing investments in this area have witnessed a higher growth rate than other renewable generation technologies.

However, due to the ongoing financial downturn, investments from Europe are anticipated to drop in 2012 and perhaps into the first half of 2013. In North America, by contrast, high investments in both solar and wind power will continue, but the largest growth area for investment is predicted to be the Asia Pacific region, as low-cost equipment manufacturers tempt the market with competitive pricing.

In the Middle East and North Africa, Abu Dhabi, Dubai, Saudi Arabia, Morocco, Algeria and Jordon, are also forecast to become popular solar power investment destinations for major market players.

The Chinese government has planned investments of around $100m to develop solar projects in 40 African nations. This will enable China to expand its production targets and bring prices down further, with an anticipated 3.6 GW of module production capacity by the end of 2012.

Globally, by the end of 2011, approximately 67.4 GW of solar PV capacity was installed worldwide (EPIA, 2011), but by 2020 this figure is expected to grow to between 101 and 138 GW (US Energy Information Administration, International Energy Outlook, 2010).

## 7.1 **Drivers for PV**

The following is a list of UK legislation with references to climate change that provide incentives or opportunities to invest in solar power:

- the Climate Change Levy and the Climate Change Agreements, which require a 50% reduction in carbon emissions from large energy users by 2025
- the Energy Act 2011, which enables a lot of the initiatives below

- Energy Market Reform (EMR), which will make it easier for small generators to enter into the market and sell surplus power. Expected to be complete by Spring 2013
- the Renewables Obligation (RO), which offers support worth two ROCs (Renewable Obligation Certificates) for each megawatt hour (MWh) of PV generated from systems sized over 5 GW. Ground-mounted solar farms up to 40 MW in size are being planned in the south of England, and are expected to be increasingly popular, with up to 1 GW expected in the next two years. Smaller installations are now supported by:
- Feed-in tariff support (see above)
- DECC has also offered discussions with developers about projects which can benefit from a proposed Feed-in Tariff with Contracts for Difference (FiT CfD) and the capacity mechanism under the EMR. Projects that could be considered would be those which might not be eligible for the Renewables Obligation or able to receive ROCs before the end of March 2017
- Green Investment Bank – to start investing mid-2013, with £3 billion overall to lend up to 2015, but unable to borrow until after 2015 at the earliest. Its priorities however will be offshore wind and waste-to-energy
- the Localism Act and National Planning Policy Framework (NPPF), allowing Neighbourhood Plans and Local Plans to favour renewable energy and sustainable development
- The Code for Sustainable Homes, which requires all new homes and non-domestic buildings to increase the use of on-site low and

net zero carbon energy generation in new buildings by 2016 in order to be zero carbon by 2016 (under Building Regulations)

- the Low Carbon Buildings Programme, which manages a number of grant schemes to cover part of the costs of installing solar thermal, solar photovoltaics, wind turbines, small-scale hydro turbines, ground and air source heat pumps, wood-fuelled boilers or pellet boilers
- Departmental Carbon Budgets, which require reductions in carbon emissions from government buildings
- Local Authority strategies, which favour reductions in emissions and have local or regional targets for renewable energy
- separate targets and standards for the devolved administrations such as, in Wales, the Low Carbon Energy Policy statement and Technical Advice Note 22 (Wales) (Sustainable Buildings (2010)), and in Scotland, a target of 20% of energy from renewable sources by 2020 and the Second National Planning Framework
- the Committee on Climate Change, the independent watchdog which was set up to ensure the UK meets its legally binding greenhouse gas reduction targets over several four-year 'budget' periods
- the Office for Renewable Energy Deployment (ORED) within the Department of Energy and Climate Change (DECC), whose job it is within government to ensure it meets its targets for the deployment of renewable energy.

Some of the above is summarised on the Government's Enabling the Transition to a Green Economy roadmap (**http://www.businesslink.**

**gov.uk/bdotg/ac'ion/detail?itemId=1096705244&type=ONEOFFPAGE&furlname=greenec'nomy&furlparam=greeneconomy&ref=http%3A//www.defra.gov.uk/environment/economy/&domain=www.businesslink.gov.uk**) and the Renewable Energy roadmap (**http://www.decc.gov.uk/en/content/cms/meeting_energy/renewable_ener/re_roadmap/re_roadmap.aspx**).

## 7.2 **UK Standards**

The Microgeneration Certification Scheme (MCS) (**http://www.microgenerationcertification.org**) provides standards and warranties for microgeneration products and installers in the case of projects that are supported by Government schemes. Only installers on this approved list can be used for such projects. It can in some circumstances be more expensive to use an installer on this list than one which is not, but this may be outweighed by the benefits. In any case, it is advisable to seek independent reports on an installer's record before committing to them.

For PV on buildings, CLG's Competent Person Schemes (as provided by section 12(5) and Schedule 2A to the Building Regulations 2000 as amended) applies. Both this and the MCS form part of the Qualifications and Credit Framework, which provides information and a simplified process for installers to identify training needs and access approved courses. The Qualifications and Credit Framework is regulated by the Office of Qualifications and Examination Regulation (Ofqual) under Part 7 of the Apprenticeship, Skills, Children and Learning Act 2009.

BRE Training (**http://www.bre.co.uk/training.jsp**) is a Government-linked organisation offering training and assessment on solar PV and connected topics.

Certification and testing in relation to a product is based on third-party verification (European Standard EN 45011 for certification and EN 17025 for product testing).

The National Renewable Energy Action Plan for the United Kingdom Article 4 of the Renewable Energy Directive 2009/28/EC contains a summary of the UK's National Renewable Energy Policy with an overview of all policies and measures to promote the use of energy from renewable resources. It also covers administrative procedures and spatial planning, technical specifications, certification of installers, electricity network operation. Under this Directive, PV is expected to have 280 MW of installed capacity (generating 240 GWh) in 2012, 490 MW (410 GWh) in 2013, rising to 730 MW (610 GWh) in 2014. This should provide confidence for investors that support will be available to fulfil these legally binding targets.

## 7.3 **Planning**

Since April 2008, PV is classed as 'permitted development' under the Town and Country Planning (General Permitted Development) Order 1995, and permission does not need to be sought. This category includes most domestic microgeneration in England, including solar PV, solar thermal, ground and water source heat pumps. Even so, planning law treats as 'development' all external installations, including solar panels, and the local Planning Authority should be informed.

It is important that the panels should not protrude more than 200 mm when installed. In the case of installations on flat roofs, listed buildings and buildings in conservation areas, World Heritage sites and designated landscape areas, the local authority should be consulted. It may be necessary to obtain approval from the Building Control office.

During 2012, the National Planning Policy Framework for England (NPPF) was introduced, which enshrines in law a 'presumption in favour of sustainable development', under which solar projects will be more favourably considered, with the default position being 'yes'.

## 7.4 **Business plans**

A business plan for a solar project should cover a 25- or 30-year period in order to compare the levelised cost of the project with other potential schemes. All system components should be included, plus grid connection costs and the cost of capital. Obtain three installers' quotes to cover the installation costs. Such costs may also include:

- scaffolding (if needed)
- roofing works or support structure
- internal work to install wiring
- connection agreement with the Distribution Network Operator (DNO)
- lightning protection
- qualified electrician's work
- a meter to monitor and record output and temperature
- hand-held display meter.

Most domestic-sized solar PV arrays are between 1.5 kWp and 3 kWp in size. A typical 2.2 kWp system will cost around £12,500.

The cost of energy in the future is unknown (for comparison), but government projected figures are generally used, even though they do not always correspond to market prices. These are obtainable from the Department of Energy and Climate Change (**http://www.decc.gov.uk/en/content/cms/about/ec_social_res/analytic_projs/analytic_projs.aspx**).

The cost savings of the system are estimated using the formula in section 6.5.1. Businesses may be able to claim financial credit for the $CO_2$ saved, which can be estimated using the information in section 6.5.2.

### 7.4.1 Implementation timescale

The project cycle for implementing renewable projects is in the order of six to 18 months. Payback begins immediately. A return on investment should be achievable within 10 years. Thereafter the project is making money.

## 7.5 **Financial support**

There are a number of finance options available besides bank finance, including Government guaranteed lending schemes, community support schemes, grants, crowd-sourcing and pre-selling output.

The Carbon Trust (**http://www.carbontrust.co.uk/**) offers leases, loans and other financing options from £1000 upwards with no maximum to all types of organisations. Payments are calculated so that they are offset by the anticipated energy savings, so that the financing option is designed to pay for itself.

Salix Finance Ltd (**http://www.salixfinance.co.uk/home.html**) is another source of loans; it is an independent, not-for-profit company, funded by The Department for Energy and Climate Change, The Welsh Assembly Government and The Scottish Government via The Carbon Trust. It exists to accelerate investment by public sector bodies in energy efficiency technologies through invest-to-save schemes.

### 7.5.1 Grants and subsidies

Information on these is available from the sources of advice in section 7.1 and section 9.

### 7.5.2 Community projects

A community (e.g. a local authority, a social enterprise, charity) will have to raise funding to cover the upfront costs for a community scheme as well as those for feasibility studies, planning, core funding and running costs, maintenance costs and monitoring (for grants). It is usually eligible for government support, although as this is changing at the time of writing it is best to seek advice from a source in section 9.

It is also possible to issue shares on a commercial basis, as evidenced by many community schemes. A not-for-profit or community enterprise would need to be set up as the legal vehicle. If the vehicle is a charity, it becomes able to apply to trusts specialising in lending to renewable energy projects.

### 7.5.3 Tax rebates

Tax relief is available on environmental goods and services. Enhanced Capital Allowances (ECAs) allow a business to improve its cash flow through accelerated tax relief on the purchase of energy-saving technologies that are specified on an Energy Technology List (ETL) which is managed by the Carbon Trust on behalf of the Government. Electricity generating technologies are not as it is about electricity saving, although solar thermal technologies are. The business is advised to consider installing energy saving equipment, however, before investing in solar electricity items on this list.

## 7.6 **Investment opportunities**

Companies, organisations and individuals in the developed world who do not wish to install solar power themselves, can still invest in it. The most popular means of achieving this is through offsetting their own carbon emissions by investing in solar and other renewable energy projects in developing countries, usually through an intermediary or broker using carbon credits. Such systems, although controversial, are independently audited and assessed to ensure that the claimed amount of carbon offset is actually saved, and that it would not have been saved without this investment.

There are two types of credits: those purchased within the UN Kyoto Protocol system as part of the Clean Development Mechanism (CDM) and Joint Implementation, and those without, called Voluntary Emissions Reductions.

### 7.6.1 The Clean Development Mechanism (CDM)

The Clean Development Mechanism (CDM) allows Annex I (developed) countries to meet part of their greenhouse gas emission caps using 'Certified Emission Reductions' from CDM emission reduction projects in developing countries. These can include PV projects.

The most appropriate method is through the UN's Programmes of Activities (**http://cdm.unfccc.int/ProgrammeOfActivities/index.html**).

These group several project activities under a single programme of activities. An example can be the distribution of solar lighting in a given area.

As of March 2012, fewer than 20 CDM registered projects were solar PV.

### 7.6.2 Joint Implementation (JI)

Joint Implementation (**http://ji.unfccc.int/JI_Projects**) applies the same principle to developed countries. Project participants from two (or more) Annex 1 Parties may jointly implement an emissions-reducing project in the territory of an Annex 1 Party, and count the resulting emission reduction units towards meeting the Kyoto target of the other involved Annex 1 Party/ies.

The sale of the resulting emission reduction credits provides an additional revenue stream for the project owners and developers. In many cases this can be equivalent to anywhere from 10% to 40% of the project investment costs. JI also enables the transfer of efficient technologies and best available practices to the host countries. For investing countries, JI helps to meet their emission targets under the Kyoto Protocol in a cost effective way.

The Netherlands, Denmark and Austria are currently among the most active buyers in JI projects. Chief selling countries include Bulgaria, Czech Republic, Romania, Poland, Russia and Ukraine, the latter two of which are rapidly increasing the number of prospective projects.

As of March 2012, there were 314 JI projects submitted for determination, with expected emission reductions of about 27 million tons of carbon dioxide equivalent per year. However, none of them are solar PV projects.

## 7.6.3 Voluntary Emissions Reductions or Verified Emissions Reductions (VERs)

These are carbon credits produced outside the Kyoto Protocol compliance regime for projects which also include wind power, hydropower, renewable biomass, forestation, and efficient stoves. This market is smaller and less liquid than the compliance market, but wider in scope.

Credits are independently verified via a Voluntary Carbon Standard (VCS) (**http://www.v-c-s.org/**), designed by the International Emissions Trading Association (IETA) (**http://www.ieta.org/**), and non-profit organizations The Climate Group (**http://www.theclimategroup.org/**) and the WWF (**http://www.wwf.org.uk/filelibrary/pdf/august06.pdf**).

The Climate Group has produced a guide for purchasing carbon offsets (**http://www.theclimategroup.org/our-news/news/2007/8/21/top-ten-tips-for-purchasing-carbon-offsets/**).

The quality of some voluntary offset schemes have been criticised recently by the Financial Services Authority (**http://www.fsa.gov.uk/consumerinformation/scamsandswindles/investment_scams/carbon_credit**), so it is important to choose a trustworthy one. It advises that before buying carbon credits, investors should check whether the credits are recorded on a registry. The three main registries for the voluntary market are APX, Caisse des Depots CDC Climat and Markit. They should also check whether the credits have been verified or accredited and whether they can be traded, that is, whether there is a demand.

## 7.6.4 Loans to installers

It is also possible to invest by offering loans to others who cannot afford the upfront costs. This is usually done through relationships with installers who act as intermediaries, marketing both their services and the loan. The package is structured to repay the loan by instalments through the feed-in tariff payments while still giving the property owner lower electricity bills than before. This is a type of leasing arrangement similar to that used to sell new cars. It offers a long-term guaranteed rate of return on the investment.

## 7.7 **Utility level developers**

There are efficiencies of scale for generating and supplying energy at a local level. In respect of solar photovoltaic power, one developer solution might be to rent and clad roofs in an area with solar panels, using government support, and sell the energy generated to local clients. The vehicle for implementing this is an Energy Service Company (ESCO).

### 7.7.1 Energy Service Companies (ESCOs)

An Energy Service Company (ESCO) is a private or social enterprise which supplies the service of energy to local clients. An organisation can negotiate to buy energy services from a local ESCO, if one is available. A developer, such as a business generating energy for its own use, may set up an ESCO to supply surplus energy it doesn't need to local clients, as part of its business model. These supply power on a local network to avoid National Grid connection fees. There are lower transmission losses and it should be cheaper. The network should be designed to minimise losses through conversion. ESCOs are incentivised to practise, and advise on, energy efficiency to minimise their costs of supplying energy.

One means of financing the setting-up of an ESCO is to enter into a fixed-price, fixed-term service agreement with local customers. This provides a future income as a loan guarantee.

CHAPTER 8

# Sources of Information

## 8.1 On official UK policy

- *Carbon price floor*, which explains this means of supporting low carbon electricity generation in the context of the government's proposed electricity market reform: **http://www.decc.gov.uk/assets/decc/11/policy-legislation/emr/2176-emr-white-paper.pdf**
- *Committee on Climate Change*, which sets out the roadmap and targets for UK action on climate change: **http://www.theccc.org.uk/**
- *Energy and emissions projections for the UK*: **http://www.decc.gov.uk/en/content/cms/about/ec_social_res/analytic_projs/en_emis_projs/en_emis_projs.aspx**
- *Feed-in Tariffs with Contract for Difference (FiT CfD)*, which explains this means of supporting low carbon electricity generation in the context of the government's proposed electricity market reform: **http://www.decc.gov.uk/assets/decc/11/policy-legislation/emr/2176-emr-white-paper.pdf**
- *The Office for Renewable Energy Deployment (ORED)*, the government body overseeing renewable energy deployment in the

UK: **http://www.decc.gov.uk/en/content/cms/meeting_energy/renewable_ener/ored/ored.aspx**

- *Ofgem*, the energy regulator: **http://www.ofgem.gov.uk/**
- *Renewable Energy Planning Database (REPD)*, tracks the progress of thousands of renewable electricity projects through planning, construction and operational phases: **https://restats.decc.gov.uk/cms/planning-database/**
- *The Renewable Energy Roadmap* (published in July 2011) includes illustrative 'central ranges' for eight key technologies - including renewable electricity technologies - and while they do not represent technology specific targets or the level of our ambition, they do show what could be deployed by 2020: **http://www.decc.gov.uk/en/content/cms/meeting_energy/renewable_ener/re_roadmap/re_roadmap.aspx**
- *Renewables Obligation*, the system which helps to support large-scale renewable energy generation: **http://www.decc.gov.uk/en/content/cms/meeting_energy/renewable_ener/renew_obs/renew_obs.aspx** and **http://www.ofgem.gov.uk/Sustainability/Environment/RenewablObl/Pages/RenewablObl.aspx**
- *Restats*, the DECC database that holds data on the status of large-scale renewable electricity generation projects, including those under construction. It is not possible to tell when projects will become operational, however the date on which generation commenced is recorded: **https://restats.decc.gov.uk/app/reporting/decc/datasheet**

- *They Work For You*: use this website to track every mention of solar energy in Parliament: **http://www.theyworkforyou.com/search/?s=solar**

# 8.2 **On solar irradiance**

## 8.2.1 MIDAS

**http://badc.nerc.ac.uk/view/badc.nerc.ac.uk__ATOM__dataent_ukmo-midas**

Land surface observations data from the Met Office station network and other worldwide stations as stored in the Met Office MIDAS database. Data are available for the period 1853 to present. The dataset comprises daily and hourly weather measurements, direct and diffuse radiation, hourly wind parameters, max and min air temperatures, cloud cover, soil temperatures, sunshine duration and radiation measurements and daily, hourly and sub-hourly rain measurements and some climatology data.

It is updated monthly. Use it to find the global horizontal irradiance (global irradiance is the total received on a horizontal surface direct from the whole sky including the sun). The method is, to find the figures for your nearest weather stations, interpolate them to get the local horizontal global figures, then to calculate the local irradiance on the tilt plane of the panels by interpolating the MIDAS irradiance and from there to work out the PV modules' efficiency.

## 8.2.2 Juice-o-meter

The Juice-o-meter (**https://solarjuice.com/**) is a UK tool based on 10 years of hourly data taken from weather stations across the UK using MIDAS, in a user-friendly fashion. Just enter a postcode, orientation and tilt angle. It also calculates system output, taking into account a 0.5% panel

**FIGURE 8.2.2. Screenshot of the Juice-o-meter.**

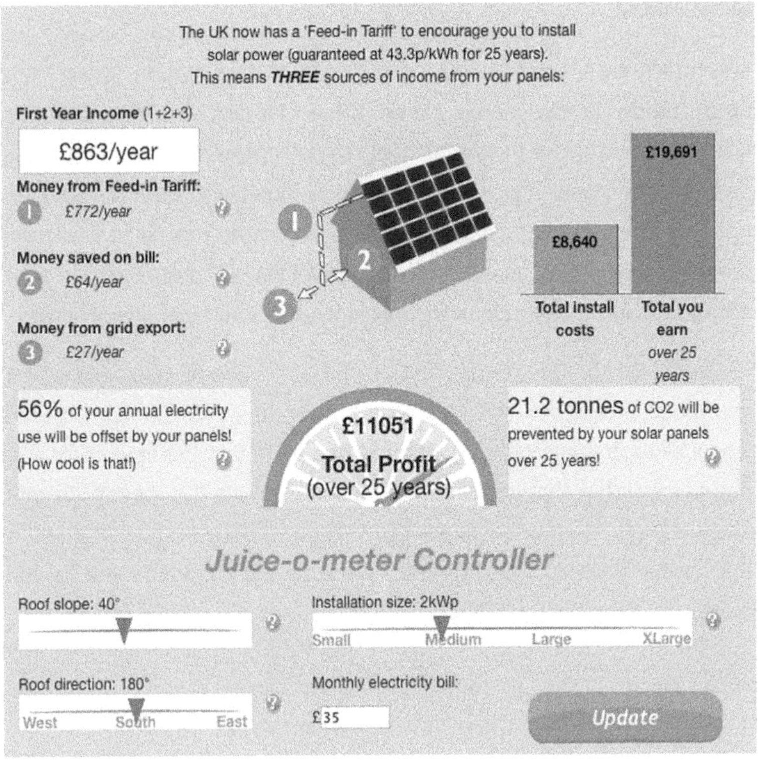

performance loss per year and a 'derate' factor to account for losses from the inverter, which is set at 85%. An installer will provide a more detailed prediction taking into account the actual panel/inverter selected.

## 8.2.3 Photovoltaic Geographical Information System (PVGIS)

The Geographical Assessment of Solar Resource and Performance of Photovoltaic Technology for Europe and Africa is at: **http://re.jrc.ec.europa.eu/pvgis/**.

Input the location, size, efficiency and other information about the proposed installation and it will calculate the average output of the system.

**FIGURE 8.2.3. Screen shot of sample PVGIS information.**

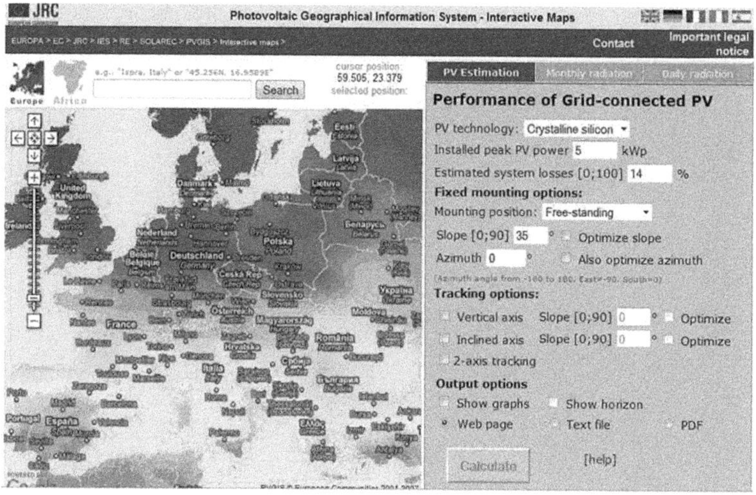

## 8.3 **Advice**

Independent advice on PV projects is available from the Energy Saving Trust (**http://www.energysavingtrust.org.uk/**) for householders and the Carbon Trust (**http://www.carbontrust.co.uk/**) for businesses and organisations. The former offers a buyer's guide to solar electricity panels on its website.

Other sources of advice are:

- The Centre for Sustainable Energy (CSE) (**http://www.cse.org.uk/**)
- BRE (**http://www.bre.co.uk/**)
- The Solar Energy Society (**http://www.uk-ises.org/**)
- The International Solar Energy Society (ISES) (**http://www.ises.org/**)
- Global Solar Council (**http://www.globalsolarcouncil.net/**), a CEO-level coalition of international companies involved in the solar photovoltaic value chain.

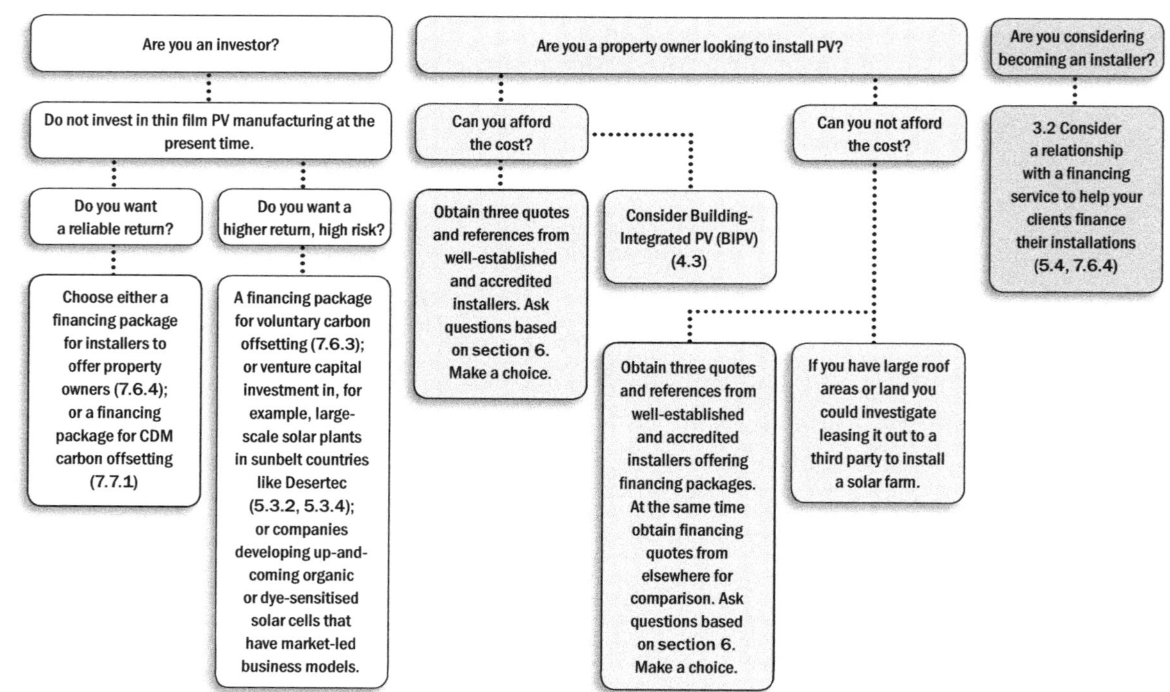
Decision Tree
Are you an investor?
Do not invest in thin film PV manufacturing at the present time.
Do you want a reliable return?
Choose either a financing package for installers to offer property owners (7.6.4); or a financing package for CDM carbon offsetting (7.7.1)
Do you want a higher return, high risk?
A financing package for voluntary carbon offsetting (7.6.3); or venture capital investment in, for example, large-scale solar plants in sunbelt countries like Desertec (5.3.2, 5.3.4); or companies developing up-and-coming organic or dye-sensitised solar cells that have market-led business models.
Are you a property owner looking to install PV?
Can you afford the cost?
Obtain three quotes and references from well-established and accredited installers. Ask questions based on section 6. Make a choice.
Consider Building-Integrated PV (BIPV) (4.3)
Can you not afford the cost?
Obtain three quotes and references from well-established and accredited installers offering financing packages. At the same time obtain financing quotes from elsewhere for comparison. Ask questions based on section 6. Make a choice.
If you have large roof areas or land you could investigate leasing it out to a third party to install a solar farm.
Are you considering becoming an installer?
3.2 Consider a relationship with a financing service to help your clients finance their installations (5.4, 7.6.4)

# About the Author

**DAVID THORPE** writes environmental material, non-fiction, fiction and journalism. He has been the News Editor of *Energy and Environmental Management* magazine since 2000. His books include the *Earthscan Expert Guide to Sustainable Home Renovation* (Earthscan, 2010) and the *Earthscan Expert Guide to Solar Technology* (Earthscan, 2011). He also writes or has written for Greenpeace, Grant Thornton, *Business Green*, Corlan Hafren (a group comprising engineering consultancy Halcrow, Arup, and KPMG), the *Guardian Online* and many others. He was Managing Editor at Centre for Alternative Technology Publications in the 1990s. His non-fiction includes an educational book for kids, *How The World Works*. He is also the author of *Hybrids* ('Essential reading for the cyberspace generation'), winner of the HarperCollins-Saga Magazine 2006 Children's Novelist competition, and has written and edited many scripts, comics and cartoon strips. He was a co-founder of the London Screenwriters Workshop, and co-wrote *The Fastest Forward* for Comic Relief, a feature film starring Jerry Hall. His career includes being the only person in the world (probably) to hold a degree in Dada and Surrealism! He also finds time to play in a band and run a media company, Cyberium. He lives in Wales and was born in Robin Hood country, Nottingham.

For Product Safety Concerns and Information please contact our EU representative GPSR@taylorandfrancis.com
Taylor & Francis Verlag GmbH, Kaufingerstraße 24, 80331 München, Germany

www.ingramcontent.com/pod-product-compliance
Ingram Content Group UK Ltd.
Pitfield, Milton Keynes, MK11 3LW, UK
UKHW040928180425
457613UK00011B/304

* 9 7 8 1 9 0 9 2 9 3 0 3 8 *